THE SEQUENCE OF LATENT TRUTHS

Wenzl Mcgowen

ISBN: 978-1-958324-52-3

CONTENTS

CHAPTER 1:
All That Is

The psychological and metaphysical interpretation of the mystical experience, includes interviews with Cognitive psychologist Donald Hoffman, NASA Physicist Tom Campbell, and inventor Federico Faggin

A few years ago I was sitting at a beach with my brother. He looked at the ocean thoughtfully and said, "Do you know the feeling of being one with all that is?"

I was irritated by his esoteric jargon, and we got in an argument about science and mysticism. At the time, I was a freshman at the New School in New York. I had been taking philosophy classes and I wanted to show off some of my new ideas.

In the 1920s, Sigmund Freud had written about the sensation my brother was talking about, referring to it as the "oceanic feeling." Freud believed that people who experienced a feeling of oneness with everything in the world were actually temporarily accessing thoughts and emotions they had as infants, before they were capable of perceiving the difference between things that were part of them and things that were separate from them. As far as Freud was concerned, these people weren't describing a genuine connection so much as they were experiencing a psychological regression.

When I tried to explain this to my brother, he just smiled and said, "So you've never experienced it, huh?"

He was right. I had never experienced anything like that. I told my brother that his mystical feelings were all in his head, that his epiphanies were nothing but electrons flying back and forth between his neurons. Back then I believed in the philosophical doctrine known as materialism or physicalism. Materialism is the philosophy that everything, including consciousness, can be reduced to physical processes.

My perspective changed a couple of years later when I signed up for a Vipassana meditation retreat. I had read some scientific studies conducted by Harvard scientists that proved that meditation can lower stress hormones and even help with chronic illnesses. Living in New York City had been quite stressful, and I figured that if meditation could help me cope with stress and anxiety, it might be worth to sign up for a retreat.

When I arrived at the retreat center, I tried my best to tolerate the spiritual atmosphere. I didn't have any interest in singing spiritual songs and listening to people talk about their past lives. However, it turned out that just a few days of silent meditation put me in a completely different mindset.

Initially, it was extremely difficult to sit for 10 hours a day and to keep my attention on my breath. My back and my legs began to hurt and all sorts of thoughts tormented me, blaming me for signing up for this. Despite the physical and emotional pain, I remained seated. Eventually my mind became quiet and my attention steady. I no longer felt the need to get up, and I often remained seated during the breaks. On the sixth or seventh day I had a breakdown — or a breakthrough, depending on how you look at it.

It wasn't complicated, and at the same time it was far too complicated to describe with words. I simply looked at a tree and couldn't hold back my tears. I felt that the same energy that went through me also went through the tree. In this state, I experienced consciousness as an all-pervasive energy that had manifested itself in various forms. I understood that my separate identity was an illusion and that I was one with all.

Within this expanded state of consciousness, I experienced a transcendental intelligence that seemed beyond anything I had previously experienced. My rational brain was trying to come up with an explanation. I was telling myself that this was all in my head, but this other form of intelligence demanded to be recognized as an awareness far beyond my current level of understanding. In fact, it told me that "the world of forms" wasn't real and that I had been identifying with an illusion. If I had been religious, I would have thought that I had an encounter with God, but instead, I was shocked and utterly confused.

This experience changed my life, because it felt more real than anything I had previously experienced. It made me question the reality of the so-called physical world. It shifted my perspective on life. I began to ask myself whether mystical experiences were more than just fabrications of the mind, as I'd previously believed. I began to wonder if perhaps they were glimpses of a deeper reality, a truth beyond the physical realm.

Before I had this experience, I would have agreed with Freud that the experience of unity with all is a psychological phenomenon, but now it seems equally plausible that I had experienced a deeper reality and that everything I thought was real is more like a dream.

Mystics from all ages have long had this type of experience. The descriptions vary, but the idea that something like "God is beyond the illusion of the physical reality" pops up in all spiritual traditions. Christians, Buddhists, Taoists, Hindus, Muslims, and shamans from around the world have said in one way or another that the supreme reality is beyond the illusion of the physical reality. You also find this idea expressed in Gnosticism and the thoughts of various Greek philosophers.

It's not too surprising that various religious and spiritual traditions share some underpinning concepts. What *is* quite surprising is that in the last couple of decades, these same ideas have also begun to emerge in physics.

Traditional physicists feel that mystical experiences should not be considered part of science. Many mystics are willing to step outside of reason and logic to explain what they experience, while physicists use mathematics and experimental evidence to carefully explore and define their findings. Mystical writings are often mixtures of vague terms and poetic phrases which aim to conceptualize otherwise indescribable experiences, while scientific writings are rigorous and detailed theories addressing observable and verifiable modifications of material interactions. It is easy for these two worlds to talk past each other, but I don't believe that science and mysticism are mutually exclusive.

The first decades of the 21st century have seen a growing movement among physicists that proposes that physical reality is actually an illusion. This *simulation hypothesis* proposes that the physical universe is not actually physical, but rather is made of information. In this model, the information that we perceive as matter is computed by a more fundamental reality that is not based on a physical substance, but on consciousness.

In physics, this is a relatively new and outrageous idea, but it is gaining popularity because materialism — the traditional belief that nothing exists except matter and its interactions — has continued to fail in explaining quantum mechanics and its apparent relationship to consciousness. The assumption that our universe is a simulation seems counterintuitive, but is better at explaining why the speed of light is fixed, how the Big Bang happened, and why large objects like planets and stars warp space.

Any simulation has a limit to how fast things can move. That limit is defined by how often the simulation updates. For example, in a video game, a character or object cannot move faster than the sample rate which determines when all pixels are updated. To create the illusion of movement, a simulation changes the colors of its pixels, which we interpret as moving objects. In our reality we also find a sample rate and pixels, which physicists refer to as quantized time and quantized space. Both make sense if we assume that we live in a simulation.[1]

Warped spacetime is another artifact that we would expect to find in a simulation. The simulation hypothesis argues that huge objects like planets and stars warp space and time because their processing load is larger than the empty space around them. It takes longer to compute their activity, which is what we experience as warped space and time.

The nature of the Big Bang is another mystery that makes more sense from the perspective of the simulation hypothesis. Instead of the belief that the entire universe was stored in an infinitely small point and then exploded suddenly, the simulation hypothesis suggests that the Big Bang was actually the launch of an iterative program.

---

[1] Tom Campbell, *My Big Toe,* (Lightning Strike Publishing 2001)

Although the simulation hypothesis explains several of the big mysteries of the universe, it's still considered an outrageous idea by most scientists, because it renders all aspects of our reality an illusion. Skeptics argue that it would be impossible to prove or disprove whether we live in a simulation. However, theoretical physicist Dr. James Gates of the University of Maryland famously claimed in 2011 that he'd found evidence of a form of computer code in the laws of physics. Gates discussed this idea with fellow physicist Neal deGrasse Tyson during a panel discussion at the 2011 Isaac Asimov Memorial Debate:

> *GATES: I've been for the last 15 years trying to answer the kinds of questions that my colleagues here have been raising. And what I've come to understand is that there are these incredible pictures that contain all of the information of a set of equations that are related to string theory. And it's even more bizarre than that, because when you then try to understand these pictures you find out that buried in them are computer codes just like the type that you find in a browser when you go to surf the web. And so I'm left with the puzzle of trying to figure out whether I live in The Matrix or not.*
>
> *TYSON: Wait – you're blowing my mind at this moment. So, you're saying – are you saying your attempt to understand the fundamental operations of nature leads you to a set of equations that are indistinguishable from the equations that drive search engines and browsers on our computers?*
>
> *GATES: Yes. That is correct. So…*
>
> *TYSON: Wait wait. I have to just be silent for a minute here. So you're saying as you dig deeper, you find computer code written in the fabric of the cosmos.*

> *GATES: Into the equations that we want to use to describe the cosmos, yes.*
>
> *TYSON: Computer code.*
>
> *GATES: Computer code. Strings of bits of ones and zeroes.*
>
> *TYSON: It's not just ... sort of resembles computer code, you're saying it is computer code.*
>
> *GATES: It's not even just computer code, it's a special kind of computer code that was invented by a scientist named Claude Shannon in the 1940s. That's what we find buried very deeply inside the equations that occur in string theory, and in general in systems that we say are super-symmetric.*

After some more discussion on the topic, Gates concluded half-jokingly: "I have in my life come to a very strange place because I never expected that the movie *The Matrix* might be an accurate representation of the place in which I live."

Entrepreneur Elon Musk and CEO of SpaceX is another strong supporter of the simulation hypothesis. Musk argues that if there ever were an advanced alien civilization, it would have had to develop an information processing system. If that information processing system evolved at the rate our computers evolve, after a couple of thousand years they would have been capable of simulating universes. If an alien race mastered such technology, it probably would have simulated millions of different universes. So if one physical universe can produce millions of simulated universes, then there is a much higher chance that we are currently living in a simulated universe as opposed to a real universe.

This train of thought still assumes that there is such a thing as a real universe from which simulated universes arise. Not all supporters of the simulation hypothesis believe that this is the case.

Donald D. Hoffman is a well-respected scientist who developed a theory based on the simulation hypothesis, but he rejects the idea of a material universe all together.

My wife and I visited him during a road trip through southern California, and we spent the day talking to him and his wife about vegan food, photography and his version of the simulation hypothesis. He brought us to a vegan restaurant south of Los Angeles and introduced us to a wide variety of meat substitutes and uniquely prepared vegetables. During the dinner we discovered that my wife and his wife both are photographers, so we found ourselves moving between a variety of topics and interests. Donald is a very kind and humble man who speaks eloquently and precisely about extremely complex ideas but also likes to listen to the passion projects of others.

Dr. Hoffman earned his doctorate in philosophy and computational psychology at MIT, was a research scientist at the laboratory of artificial intelligence at MIT, spoke at the TED conference, and now is a professor at UCI. He has written books on vision and perception and proposes now that the reality we experience is the user interface of a "conscious agent," Hoffman's term for a unit of awareness that perceives and makes choices.

During our meeting we talked a lot about what Hoffman calls MUI theory. MUI stands for *multimodal user interface*. He told me that when we play tennis online, it appears that there is a ball, but really the ball is a visual representation for the complex reality of transistors that facilitate the data flow within the computer. He says that our experience in time and space is no different. The reality that provides us with the experience of a physical world is a network of conscious agents that have nothing in common with the physical objects we experience.

Hoffman believes that each conscious agent has its own data stream, and therefore we all experience a different reality. For example, when I hand you a spoon, we both assume that we see the same spoon, but MUI theory suggests that we each interact with a different spoon, which isn't a physical object, but a symbol in the data stream we are receiving.

When I asked why we're still able to share our experiences with others, Hoffman said that our data streams are synchronized. That is why we can agree on one reality, but the reality we agree on is only an experience, not an external real world.

Hoffman believes that our user interface, which is our experienced reality, has evolved to facilitate the evolution and data exchange of a system that exists beyond space and time. Every action we take in this world is a real data exchange, but the physical objects we perceive are only the icons in our data stream. In the same way, when you move a file on your desktop, the user interface of the computer hides the complexity of the processes occurring within the hard drive of the computer. Moving the file still changes the data flow within the computer, so the illusion is a helpful tool. If you had to deal with the complexity of the computer's transistors every time you wrote an email, you would never get any work done. The interface is efficient because it hides the complexity of what really happens.

In the same way, evolution has rewarded those conscious agents whose interface has allowed them to navigate the information system effectively, and this perceptual evolution gave rise to what we call time, space and the experience of physical objects. However, the reality we perceive is a simplified model of a more fundamental and much more complex world.

Tom Campbell is another scientist who developed a theory based on the simulation hypothesis, which he refers to as *MBT* (short for "My Big Theory of Everything"). He worked for U.S. Army technical intelligence, then moved into research and development of defensive missile systems, and most recently worked for NASA within the Ares 1 program.

I first visited him at his hilltop home in Huntington, Alabama, and have had several follow-up interviews with him since. Campbell, an avuncular, white-bearded 70-something, speaks very calmly about subjects that dramatically derail the Western ideas of what reality is.

According to Campbell, the simulation hypothesis isn't just a hypothesis — it's the nature of his life. Campbell claims that he has gained the ability to disconnect from the data stream of our physical universe and browse through other dimensions with his "nonphysical body." I was standing next to him on his wooden terrace looking at one of NASA's compounds when he said, "Learning to live in multiple realities is like learning to drive a car. When you are a child the idea seems unfathomable, and later on you don't even think about it."

Campbell said he learned this skill in the 1970s, while he was working at the Monroe Institute researching out-of-body experiences. He told me his story with a very calm and deep voice; his demeanor was always calm, kind, and slightly uninterested, no matter whether the topic was quantum physics, living in multiple realities, or Indian food. I visited him while I was on tour with my band Moon Hooch. Our drummer James likes to cook Indian food, so he made us dinner while Tom explained his theories.

Tom Campbell spends his waking time devising quantum experiments to investigate the simulation hypothesis, but claims that he doesn't actually sleep at night. Rather, he says that he's awake in other dimensions — or other reality simulations, as he refers to them. He claims to work there with nonphysical beings, some of whom have left their bodies elsewhere and some of whom don't have a body at all anymore.

According to Campbell, the laws of time and space are but one creation of consciousness and are designed to facilitate particular types of interactions. When you leave your body, you can interact with many more realities and lifeforms that are all on different evolutionary paths, in different simulations.

Campbell is indeed a strange combination: a well-respected physicist who also claims to be living in multiple realities. It is hard for the Western mind to accept such a combination, as generally we consider anyone who claims to be living in multiple realities to be delusional. But when the experience of multiple realities positively affects your career, then what are we supposed to say? Are most people living in a very limited state of consciousness, or have people like Campbell found a way to make their insanity work for themselves?

Campbell told me that when he was working for NASA, he often had to solve extremely complex problems in which there were thousands of variables that all affected each other. He said that he would work through these problems by closing his eyes and leaving his body. While in this state, he said, he could create visual representations for the whole system and come up with solutions for problems that would have otherwise taken him days or weeks.

Regardless of whether Campbell was actually capable of leaving his physical body, it was clear to me that he had an unusually honed focus. When I asked him how he'd developed such focus, he said, "Meditation." After talking to him for a while, I began to understand that he wasn't just talking about one life. He said that in a past life, he was a yogi in the Himalayas, where he meditated all day every day and eventually decided to reincarnate as a scientist. "In each life," Campbell told me, "you acquire skills that you then can use in the next."

Tom Campbell is a very far out guy, but it is difficult to write him off as crazy; he is an accomplished physicist and is also a very kind and loving person. There was nothing crazy about him, except that he claims to be experiencing a much broader reality than most people do. For example, he told me that he already knew his grandchildren before they were born, claiming to have perceived them as nonphysical beings before they entered the physical reality simulation. Either Tom Campbell is delusional, or he is operating on a level of awareness that most humans have never experienced.

Although there are many scientists who have built theories on the assumptions of the simulation hypothesis, the scientific community is far from agreement on the topic. Most scientists think that the simulation hypothesis is unprovable and irrelevant, and that the idea of an intelligence that resides outside of the illusion can't be part of a scientific hypothesis. However, since the dawn of the 21st century, more and more researchers have been willing to acknowledge that there are compelling parallels between the simulation-hypothesis information-based worldview and the subjective experiences of mystics and spiritual seekers.

During my investigations on this topic, I was fortunate enough to schedule an interview with Federico Faggin, an Italian physicist who invented the silicon microchip and the touch screen, the two inventions that made the smartphone possible. We met at a restaurant in Palo Alto, California. Like Campbell, Faggin is in his mid-70s, and had an infectious laugh he often used throughout our meeting.

We talked for awhile about his career as a physicist, but what captured my attention the most was an experience he had when he was 49 years old. He was with his family at his ski house in Lake Tahoe; after a day on the slopes he tried to go to sleep, but instead he saw a beam of light shoot out of his heart.

"The light wasn't just visual," Faggin told me in his thick Italian accent, "it was the most overwhelming experience of love and I knew in that moment that everything was made out of this love." He continued to tell me how his sense of his body disappeared as he became this light and had the feeling of merging with all that is.

As soon as he said this, I knew that Faggin was talking about a spontaneous mystical experience. We spent the rest of our time together talking about many more mystical experiences he had and how those experiences made him expand his work in physics. While designing the gadgets of modern society, he quietly developed a framework to make sense of reality and consciousness. What he came up with strongly resembles Tom Campbell and Donald D. Hoffman's version of the simulation hypothesis. Faggin departs somewhat from Campbell and Hoffman in that he thinks quantum bits, rather than binary bits, are the foundation of the data structure of our universe. But the ontology is the same: consciousness creates experiential realities that appear physical. Or, in other words, matter is not matter; matter is *information*.

It would have been inconceivable to propose that reality is information before humanity witnessed the exponential growth of digital information processing. Fifty years ago, computers barely had the power to run a game of *Pong*; today, computers are able to process virtual realities that are realistic and interactive. It is no longer a far-fetched idea that virtual realities have the potential to evolve life.

In fact, at this very moment, there is life evolving in a human-made virtual reality called Avida. The project is under active development by Charles Ofria at the Digital Evolution Lab at Michigan State University. The digital organisms that evolve in this simulation interact with each other, reproduce, mutate, and share virtually all attributes of what we consider life. The project raises the question of what life is and if a program can be considered alive if it reproduces, mutates and interacts with its environment. But besides raising questions about the definition of life, Avida demonstrates that evolution can also function within a virtual reality. It makes one wonder if we are the first organism to create a virtual reality that evolved life or if we are perhaps the product of some other intelligence that already has explored these possibilities.

Imagine an information processing system that is far more evolved than our computers. Could it be that our universe is the product of such an information system? Could it be that religious and mystical experiences occur when consciousness momentarily frees itself from the simulation — when it sees beyond the interface of time and space?

CHAPTER 2:

Everywhere and Nowhere

The mystery of quantum mechanics and remote viewing, includes interviews with Helane Wahbeh from the Institute of Notice Sciences, and Brenda Dunne from the Princton Engineering Anomalies Research Lab

On a subatomic level, particles don't exist as solid objects; they are "probability waves" and only become solid particles when they are observed or measured. The term "probability waves" refers to a particle that is nowhere and everywhere, but is more likely to be found at certain places. A particle that is in this state acts as if it is in multiple places at once and completely disregards the materialistic beliefs of Newton and Einstein. As soon as that particle is "observed," the probability wave "collapses" and behaves like a solid particle interacting with time and space like a solid piece of matter.

The famous double-slit experiment, first performed in 1801 by researcher Thomas Young, has repeatedly confirmed this odd behavior of subatomic particles. In the double-slit experiment, a photon emitter was used to shoot photons (particles of light) at a light-sensitive film. Between the emitter and the film was a plate with two slits. After several photons had been fired, an interference pattern emerged on the film. The pattern was comprised of roughly a dozen stripes that were defined in the middle and gradually faded out to the left and right.

This is the same pattern you would get when a wave enters two slits simultaneously and interferes with itself on the other side of the slits. To the surprise of the observers, this interference pattern appeared even if the photon emitter shot individual photons. But how could a single photon interfere with itself — unless it was somehow in two places at once?

In another version of the experiment, measurement devices were attached next to the two slits. As soon as researchers were able to determine through which slit the photon traveled, the interference pattern disappeared, and the impact of the photons created two spots on the film, indicating that single particles traveled through one of the two slits. At first, researchers thought the measurement devices must have been affecting the particles. But then scientists created other experiments that did not require measurement devices that could interfere with the photon's path.

The Delayed Choice Quantum Eraser experiment uses two beam splitters, two photon paths, and two mirrors. One photon is shot against the first beam splitter, giving the photon the option of two paths. A beam splitter lets half the light pass through and reflects the other half. Therefore, photons go 50% on path A and 50% on path B.

The two mirrors bring the two paths together and lead them to a second beam splitter. Then the photon once again has the option to go on two different paths. If it gets reflected by the beam splitter, it ends up in detector A; if it goes through, it ends up in detector B.

The effect of this setup is that the scientists have no way of knowing which path a single photon followed. If the photon gets reflected by the first beam splitter, it goes on path A and gets brought to the second beam splitter, where it either gets reflected or passes through and then hits one of the two detectors. Since the same is true for path B, the experimenter cannot know through which path it traveled.

After firing a large number of individual photons, scientists found interference patterns on both detectors, indicating that the photons traveled as probability waves, going through both paths at the same time and interfering with each other.

Armed with this information, the scientists removed the second beam splitter and brought path A and B directly to two detectors. Under this setup, the scientists could know which path the photon took. This caused the photons to travel as individual particles, creating individual spots on each detector. In other words, the scientists' observation or ability to know somehow caused the photons to act differently!

In 1978 physicist John Wheeler proposed a modification to this experiment. He had the idea to insert the second beam splitter after the photon had already entered path A or B. Based on common sense, the photon would have had to enter either as a wave or as a particle. So, what happens when we change the part of the experiment that gives us the ability to know which path the photon took?

In 2015, Associate Professor Andrew Truscott and Ph.D. student Roman Khakimov of the Australian National University performed another version of the experiment. Instead of inserting a physical beam-splitter after the photon entered the contraption, they created a beam-splitter with counter-propagating laser beams which they could turn on or off. This allowed them to make the choice after the photon should have made its choice. Their experiment confirmed the "weirdness" of the quantum world. They even performed the experiment with atoms, demonstrating that isolated atoms also act the same way. They are probability waves when we are not looking and they are particles when we are, even though the experimenters made their choice after the particle entered the intersection of path A and B. So even if the experimenter takes away the beam-splitter while the atom is traveling through both paths at the same time, the atom changes its state and becomes a particle.

In Truscott's words: "It proves that measurement is everything. At the quantum level, reality does not exist if you are not looking at it."

When I found out about these experiments and theories, it gave me a way to think about my mystical experiences. I still was unable to believe in religion, but the fact that science is constantly demonstrating that reality is not what we think it is allowed me to open my mind to new possibilities. While touring through Europe and America with Moon Hooch, playing over two hundred shows a year, I was gripped by insatiable curiosity. When I wasn't playing my instruments, I was researching quantum physics and the strange territories it revealed, and in doing so I found out about the Institute of Noetic Sciences, a research organization in California.

During one of our West Coast tours I was able to schedule an interview with the Institute's Director of Research, Helane Wahbeh. I was a little intimidated by her title, but Helane turned out to be very easy to talk to. She was soft spoken and listened very attentively to my questions. We sat at a wooden table in the backyard of the dining hall and talked about the purpose of their research.

Wahbeh told me that the Institute of Noetic Sciences ("noetic" means "relating to mental or intellectual activity") started in 1973 when Edgar Mitchell had a mystical experience in outer space. "He was coming home from the moon and he was in the shuttle," Wahbeh said. "He was looking at earth and he had this incredible experience where he felt that he was connected to everything, the stars, the shuttle, the Earth, and everything else." This experience inspired Edgar

Mitchell to explore the nature of consciousness and the possibility that everything is connected through an all-pervasive field of awareness.

I already knew at this point how powerfully mystical experiences can alter one's life path, so I wasn't surprised that one had resulted in an entire research institute. I talked to Helane about some of their most recent experiments and she told me they set up a double-slit experiment designed to measure the direct impact of observation. Test subjects were asked to sit two meters from the machine and listen to instructions. There were two types of instructions: "Please influence the beam now" and "You may now relax." The scientists analyzed the interference pattern during those two periods and were able to detect a change whenever the test subjects directed their attention towards the double slit machine.

Dean Radin, one of the scientists who worked on this experiment, said that people who had trained their focus had a stronger impact on the interference pattern. From Radin's paper:

> *Variables including temperature, vibration, and signal drift were also tested, and no spurious influences were identified. By contrast, factors associated with consciousness, such as meditation experience, electrocortical markers of focused attention, and psychological factors including openness and absorption, significantly correlated in predicted ways with perturbations in the double-slit interference pattern.*

I know — all of this sounds crazy. But the Institute of Noetic Sciences isn't the only research facility that has confirmed that intention can affect material processes from a distance. Robert Jahn and Brenda Dunne, founders of Princeton's Engineering Anomalies Research Lab, came to a similar conclusion through their work.

I interviewed Dunne in Princeton, New Jersey. We sat in her living room and talked for a couple of hours. She was 75 at the time of our meeting. Her mind was sharp and she spoke clearly and eloquently. She told me that it was more than three decades ago that she started exploring the relationship between intention and probability.

Dunne's interest in this research field was ignited by the groundbreaking remote viewing studies of Stanford University. The term *remote viewing* refers to a person's ability to see events that happen too far away to be physically sensed — or even to see events that happen in the future. Stanford's research inspired many psychologists and physicists to investigate the phenomena independently. Dunne told me that she was folding laundry one day when she decided to attempt to visualize what her friend would be doing at 5 pm. She instantly had a vision of her friend walking in a forest.

Dunne finished folding her laundry and at 4:45 the doorbell rang. It was her friend, who said: "It is a nice day, would you like to take a walk in the forest?"

"My jaw dropped to the floor," Dunne told me, "and after I picked it back up we went on a walk."

This wasn't exactly scientific evidence, but it was a powerful personal experience that changed the course of Dunne's life, opening her to the possibility that the mind is more powerful than she previously thought.

Her colleague Robert Jahn, on the other hand, did not believe that consciousness had any nonphysical properties. He was a plasma physicist, Professor of Aerospace Science, and Dean of Engineering at Princeton University. Somehow Brenda managed to convince Robert to give this type of research the benefit of the doubt. Together they founded the Princeton Engineering Anomalies Research Lab and set out to explore the relationship between mind and matter.

From 1979 to 2007, Dunne and Jahn conducted experiments with random number generators that suggested the human mind has the ability to influence events at a distance. Their random number generators used the random fluctuations of positive and negative charges in electrical noise to create random sets of "heads" and "tails." During their experiments, a test subject would sit in front of the machine and try to influence it to produce more heads than tails, then try to influence it to create more tails than heads, then decide not to influence the machine at all.

After 2.5 million electronic coin throws and 30 years of experimenting, Jahn and Dunne's body of research continuously demonstrated that human intention affected the behavior of their machines. No fewer than 68 independent investigations replicated their results later on.

This sounds unbelievable, but it's just the tip of the iceberg. While other scientists were running these intention-sensitive random number generators, they not only confirmed that our will has a direct influence on their outcomes, but they also discovered that events which synchronized the consciousness of humanity affected the machines.

The Global Consciousness Project is a research group that continued the work of Brenda Dunne and Robert Jahn, specifically measuring the effects global events had on their random number generators. According to the GCP, the 9/11 terrorist attacks caused random number generators around the world to produce strings of one number. Four hours before the planes crashed there was a major spike in non-random activity, another when each plane hit, and then again when the towers collapsed. Over the following two days, random number generators continued to show deviations from randomness and then went back to normal.

The GCP has observed several more instances that showed a correlation between global events and deviations of randomness. When the results of O.J. Simpson's trial were announced on television in 1995, a similar effect was recorded. Other global events that affected the machines were the Indian Ocean tsunami of 2004 (which killed over 225,000 people), the bombing of the American Embassies in Africa in 1998, and the death of Princess Diana in 1987. Even New Year's celebrations and soccer's World Cup appear in GCP's data as fluctuations in randomness.[2]

After observing the effect these global events had on their random number generators, GCP founder Roger Nelson wrote:

> *When human consciousness becomes coherent, the behavior of random systems may change. Random Number Generators based on quantum tunneling produce completely unpredictable sequences of zeros and ones. But when a great event influences the feelings of millions of people, our network of RNGs becomes subtly structured. We calculate one in a trillion odds that the effect is due to chance. The evidence suggests an emerging noosphere or the unifying field of consciousness described by sages in all cultures.*

When I asked Brenda Dunne why these findings haven't changed the world yet, she said, "It is hard for people to change their beliefs." She told me that even the most respected scientists will reject a fact if it threatens to overthrow their worldview. Brenda said that she had to face an incredible amount of resistance throughout her career,

---

2 Lynne McTaggart, *The Field*, New Age Publishing Ltd. 2009.

but she understood that it wasn't rational when one of her colleges screamed at her: "If your research is correct, everything I have done in my life was wrong!"

I also talked to Helena Wahbeh about this topic. Wahbeh told me that she has stopped reacting to the resistance of materialists. "Within the scientific community," she said, "there is a spectrum of belief." She explained that there are the die-hard materialists and the die-hard non-materialists. The former group will not alter their world-view no matter how much evidence you show them, and the latter group already believes that consciousness isn't produced within time and space.

Interestingly, though, she added that the people who benefit most from their research are those who are frightened by mystical experiences because they contradict the materialistic beliefs they have inherited. Wahbeh said that their research shows that science isn't necessarily in opposition to mystical experiences, because the idea that consciousness is all-pervasive and interconnected is not only better at explaining quantum physics, it also reaffirms the reality one can encounter through a mystical experience.

This made me wonder if the feeling of overwhelming love that is often described as part of mystical experiences might also have a scientific basis.

CHAPTER 3:

Harmony

Subatomic coherence in living systems and the relationship between health and order, the behavior of subatomic particles and the idea of freewill

On a cellular level, each one of us is not an individual, but a collaborative social system. Each human body is comprised of over 37 trillion cells, living in harmony with each other and sharing their abilities. If our cells were not collaborating, our nervous systems would not work, and therefore neither would our bodies. We exist as human beings because our cells synchronize their activity and maintain the organization of our bodies.

If we zoom in even further, we find that cells are made of molecules, atoms, and ultimately subatomic particles. These individual particles form larger organized structures or coherent wholes. Together, subatomic particles form atoms; atoms form molecules; molecules build cells; and cells build multicellular organisms. In this reality, we see an evolution towards ever-greater complexity and organization, but why this is the case isn't entirely clear.

The doctrine of materialism states that matter moves towards low-energy states. This means that matter tends to find a configuration in which the movement of all particles reaches maximum synchronization in the given circumstances. In the simplest terms, this means that matter organizes itself. For example, when you put certain minerals together, they eventually begin to rearrange themselves and form crystals. Crystals are repetitive and highly organized atomic and molecular patterns. These self-organizing principles can be observed in any physical system.

The systems we call atoms, molecules, cells, and multicellular organisms can be conceived as expressions of probabilistic systems — systems built on probability that create coherent wholes through harmonious resonance. For example, atoms can only share electrons when there is a very high degree of order. If you think of the orbit of an electron as an oscillation (a movement back and forth at a regular speed), then the repetitive interaction of different oscillations becomes harmonious resonance. "Harmony" and "resonance" are both terms used in music, but they apply to anything that is a result of vibration. In the words of Nikola Tesla: "If you want to find the secret of the universe, think in terms of energy, frequency, and vibration."

All things that work together have resonance with each other; this means that their vibrations or repetitive movements are synchronized. When subatomic particles are sufficiently synchronized we call them atoms, and the same patterns repeat outward. When atoms are synchronized, they bond together and form molecules, which through their process of synchronization form cells and multicellular beings. All are synchronized patterns built on top of synchronized patterns or complex systems that resonate with each other or act in harmony. Each unit, whether an atom, a molecule, a cell, or a multicellular organism, is a unit or a coherent whole, because the processes within it are synchronized.

When you look at this from the cellular perspective, you realize that evolution isn't exactly survival of the fittest, it is survival of the fittest *system* — the system that collaborates the best. For example, the strongest bird that managed to pass on its DNA was a bird that was healthy, which means that there was order on the cellular level. This means that its cells worked well together. An organism with cells that don't collaborate efficiently is diseased. We can't be healthy if our immune systems attack our muscle cells or if our hearts and brains don't communicate properly. If the communication breaks down between the organs or subsystems of our body, we either die or get ill. The more organization a system or organism can create, the more efficiently it communicates and utilizes resources.

So what is the principle that creates organization in a complex system? What creates order within the atom, within the molecule, within the cell, and within the multicellular system that we call an organism?

The definition of a complex system is "any system that features a larger number of interactive components," but something needs to be driving the interaction of the components. That is usually where most scientists scratch their heads. No one really knows what's driving the interactions of particles, molecules, or cells.

All scientists agree that certain laws are guiding the interactions, but what creates the order we find on all levels of reality? The laws of physics only describe the framework within which particles interact, but *why* they create ordered systems isn't explained. How the orbits of different electrons are synchronized is a mystery.

A good friend of mine does laser shows for bands. I met him when he was doing lasers for Coldplay. My band played a few songs before and after the different acts, and as I exited the TV studio, I handed him my band's CD. Since then we have built a relationship based on our mutual fascination with music, lasers, and life.

One night we were talking about electricity at a Greek restaurant in Manhattan, and he said, "Wenzl, do you know what electricity is? We have gotten very good at manipulating it, but we don't really know what it is."

I always thought electricity was a form of energy created by charged particles, and that is true, but what drives a charged particle? Its charge and the charges around it, at the most basic level, but after that, we can only say it is driven by probability. The same is true for anything else, because everything is made of subatomic particles which are probability waves. Yet these probabilistic waves somehow organize themselves to form coherent wholes. How can probability organize information so effectively that larger systems like us humans can exist?

Cells organize themselves through collaboration. Could it be that subatomic particles also arrange themselves through collaboration? Perhaps collaboration is the same thing as coherence and harmonic resonance.

Fritz Popp, the founder of the International Institute of Biophysics, discovered that living systems use coherent photons to communicate. Popp was working with a carcinogenic substance and discovered that it scrambled light waves. He thought that the toxicity of the material could be a result of this property, and further wondered if living systems might be using photons for internal communication. To test this hypothesis, he built a darkroom and used extremely light-sensitive equipment to see if biological systems emit photons. He found that all living systems emit coherent photons. This means that, somehow, these photons can synchronize their probability waves.

Popp also noticed a correlation between the coherence of the emitted photons and the health of an organism. Unhealthy organisms emitted more photons with less coherence. This means if there is less organization on the subatomic level, there is also less organization on the cellular level, which is really what bad health is. According to Popp, this is the case because the photons of healthy organisms are part of a coherent field, or a vast network of synchronized photons, while the communication within this field has broken down in unhealthy organisms.

Dr. Konstantin Korotkov, the founder of Bio-Well, used this hypothesis to build his GDV camera, a device that can measure the photon emissions in a finger and analyze the coherence of the "human energy field" to assess a person's health. Dr. Korotkov used the phrase "human energy field" to describe light particles or photons with probability waves that have lined up to the point where trillions of particles become practically indistinguishable from each other. Fritz Popp's devices could only detect photons that had left this field. It also explains why healthy organisms emit fewer photons than unhealthy organisms. The field in healthy organisms has more coherence, and therefore fewer photons escape it; they all vibrate in harmony with each other like a fine-tuned orchestra.

Popp is not the only researcher who recognized the correlation between coherence and well-being. German scientist Martin Schöne approached the subject from a different angle but came to a similar conclusion. He discovered that states of consciousness that we experience as bliss can be measured as coherent electromagnetic waves within the brain.

I found out about his work because my brother sent me one of his YouTube videos. I was instantly fascinated by Schöne's research but didn't write to him until several years later, when I was on tour and was reminded of his studies in an unexpected way. I was doing a soundcheck with my band somewhere in Boston and as I saw the low tones create resonance images in the glass of water placed on a speaker, I remembered Martin Schöne. I wrote him an email that day and visited him the following summer.

Dr. Schöne was in his 40s when we met; a tall, handsome man with a contagious smile that continued to shine even while discussing extremely complex ideas. His long sentences effortlessly wove together quantum physics, neuroscience, Taoist philosophy, and the importance of art. Schöne, his partner Christina, my bandmate Mike and I spent the evening drinking German beer in Schöne's small backyard in Berlin, which was overgrown with vines and bushes and featured a large table next to a red brick wall. The four of us clanked our bottles together for some time, talking about science and the mystical territories that contemporary researchers are forced to enter.

Before Mike and I left, Dr. Schöne asked us if we would like to come back the next day and get our brains measured. I was very excited — and eventually very disappointed, for reasons that we'll come back to later. For now, let's focus on exactly what Schöne meant by brain measuring.

Martin uses a device that he named the Brain Avatar. The Brain Avatar uses the electromagnetic waves the brain emits to create an image. The machine does this by converting brainwaves into sound waves, then uses those sound waves to make water vibrate, and finally projects the moving image onto a screen. It looks a bit like the patterns that emerge when you put a glass of water on a loudspeaker, except that the Avatar shines light through the water and projects it on a screen.

Martin told me that the images the Brain Avatar produced showed a correlation between balanced states of consciousness and geometric symmetry. For example, the brain waves of a Buddhist monk resulted in a beautiful geometric image reminiscent of the colors and patterns you might see in an ancient church's stained-glass windows. On the other hand, the brain waves of a child with ADHD created a continually changing image with no geometric symmetries. This is the case because the brain of a monk has more synchronization, which means that different parts of the brain produce electromagnetic waves that have a harmonious relationship. Waves that have a harmonious relationship line up and create an orderly interference pattern. The brain waves of a stressed person have much less coherence and don't create an orderly interference pattern.

Imagine throwing a bunch of pebbles in a lake and watching how the rings interact with each other. The pattern will be quite chaotic, but what if you could calculate the size of the lake, the speed of the waves, and the arc of the stones so that your throws would cause the waves to line up in a consistent pattern? That would probably be impossible to do at a lake, but a balanced brain does just that. When you are in the state of zen or thoughtless bliss, all impulses in your brain line up nearly perfectly.

Dr. Schöne told me that people who experience bliss or "cosmic love" during his tests produce a coherent resonance image because the different parts of their brain are extremely synchronized. This means that the lower brain waves have overtones that line up with the fundamental pitch on a regular basis, just like two musical notes that are in tune with each other.

When two notes are in tune with each other, it means that their waves line up after an even number of oscillations. For example, if C and G are in tune with each other, the pressure wave that we call C goes up and down two times while the wave that we call G goes up and down three times. The waves line up on a regular basis and form an orderly interference pattern. If you took those two pitches and put them into the Brain Avatar, you would see a coherent resonance pattern, because as those waves bounce back and forth in the water, they always maintain this harmonious relationship of 2/3. This would not be the case if these waves had an irregular relationship, or, in other words, were out of tune. Just like well-tuned instruments can produce enjoyable music, a brain that's tuned through meditation produces emotions of bliss and love, which are measurable as electromagnetic coherence.

A year after I visited Dr. Schöne in Berlin, his videos disappeared from the internet, and he stopped replying to my emails. I later found out that he died and his partner got into patent law complications which forced her to take down some of the research. But his research left a deep impression on me. Meeting Martin Schöne inspired me to keep exploring the idea that bliss, love, compassion, or heightened positive emotions, result in coherent brain waves.

I came across the research of Fritz Popp much later, but I instantly realized that he was describing the same pattern at a different level. Fritz Popp did not measure electromagnetic waves. He discovered the correlation between health and coherence on the subatomic level. But if you zoom out, you see the same pattern.

Synchronization is not just necessary for producing happy, healthy and loving human beings. The very essence of reality also needs synchronization. If you analyze any system, whether it is the cells of a plant, the communication between different brain parts, or even the atoms in stones, you will always find the evolution towards synchronized patterns.

What is the force that creates order in the universe? No scientists can answer this with certainty. It is a question that is on the border of science and philosophy, as the answer might go beyond three dimensions, or beyond a verifiable theory.

Tom Campbell believes this force is consciousness and its innate drive to become love. On its surface, this idea might seem vague and cloyingly New Agey, but as long as we clearly define what we mean by "love," we can use this term precisely.

It's important here to make a distinction between the *feeling* of love and the *principle* of love. The principle of love is the *will to benefit others*. It is descriptive of a relationship between will-based units, which means that anything that can form interactive relationships can become a manifestation of this principle. Anything that has free will and relates to other free will units can develop a love-based relationship — a relationship based on the intention to benefit others or the will to create harmony together.

The feeling of love is the human manifestation of this principle. Human beings have evolved an expression of this principle because mutually beneficial relationships gave tribes an evolutionary benefit through social organization. When you treat others with love and care, you create coherence or order in your social environment. Love makes us want to benefit others and makes us want to work together. That in turn creates organization, and organization is rewarded by evolution since organized units utilize energy more efficiently.

Donald D. Hoffman's MUI theory might help us at this point. He proposes that everything we see is an interface that represents the activity of conscious agents. Even subatomic particles are like icons on your desktop, but the relationships they form are real, in the sense that information is being organized.

Could it be that the organization we find on the subatomic level is a result of the will to create harmony together?

When we look at the interface of reality, we find coherent structures in matter, but the conscious agents that arranged these information structures might be experiencing love. This means that multiple conscious agents have mutually beneficial relationships and are attracted towards these relationships through a subjective experience. The result of these relationships is a coherent quantum system, which we experience as solid matter. So when we hit a wall with our toe, our interface tells us that we have come in contact with a highly organized and stable network of conscious agents. Such agents aren't going to change their relationships based on our will unless we pick up a hammer and hit the wall. This would leverage other networks of conscious agents that can have an impact on the structure of the wall.

This path of exploration can quickly become very anthropomorphic. Broadening subjective terms such as "love" and "will" is usually looked down upon by Western science. However, such disdain might be the remains of a romantic fantasy. In striving for objectivity, materialistic scientists tend to remove themselves from their experiments and theories, never contemplating that their inner world might be a continuation of principles that are present in every aspect of existence. The dawn of quantum physics challenged this attitude, as researchers were confronted with evidence that their observations themselves affected the experiments they conduct. In other words, their subjective experience was proven to be part of the objective world they tried to investigate. It isn't a major leap to propose that aspects of our subjective experience are also part of the so-called objective reality.

Our subjective experience doesn't need to be something that we try to ignore while contemplating scientific truths. It would be foolish to push directly experienced truths aside and hope that we will get around to explaining them after we are done exploring a world that we assume is real, objective, and made out of matter. As we have seen, quantum physics has exposed major problems with this perception of the world. Perhaps our inner world can point us towards a new model of reality.

Although the "external" world might just be a representation of the relationships between conscious agents, we still can use experiments to explore this hypothesis. For example, we could ask, "Is there any observable evidence that the interactions of subatomic particles are driven by a conscious agent who has free will?" If this is the case, we might be able to set up an experiment in which a particle's ability to decide is mathematically proven.

Most physicists agree that particles act based on objective probability, but not many physicists would say that objective probability is free will. Let's think about this for a moment. They call it "objective probability" because there isn't more information to be gained, which could be used to predict the future. This means that particles behave spontaneously.

Einstein had a huge problem with these interpretations of quantum physics, summed up in his famous quote that "God doesn't play dice." But as more and more scientists conduct experiments like the ones we've discussed, it's gradually become more widely accepted that perhaps God *does* like to play dice. Dice is a fun and unpredictable game, after all.

Usually, we make sense of the world by looking at a situation and analyzing how events in time and space brought the present moment into existence, but this way of thinking fails at the quantum level. Particles don't act exclusively based on cause and effect; they also behave unpredictably. They are likely to appear at certain places, and this likelihood is what we refer to as their probability wave. This uncertainty is part of their fundamental nature — and, perhaps, proof that not only mortals enjoy gambling.

The probability waves of subatomic particles could be considered "free will" as long as we define free will as a force that guides the spontaneous unfolding of the future.

Under this definition, two Princeton mathematicians, John H. Conway and Simon B. Kochen, claim to have proven that subatomic particles have free will. Their Free Will Theorem was published in 2011 by the Cambridge University Press, and many mathematicians and physicists recognized it as a significant achievement. Their paper is extremely technical, but the essence of it is that they worked with a particle that, when measured, always had a spin of either one or zero. They set up an experiment that forced the particle into a logical paradox and were able to prove mathematically that the spin of the particle is not a function of the past, which means that it decides its spin when measured. The outcome is not much different than the other experiments I discussed; the only difference is that they used the term "free will" instead of "objective probability" and proved mathematically that the spin of the particle results spontaneously.

Whether we call it objective probability or free will doesn't really matter. Ultimately, we're talking about a spontaneous force that guides the unfolding of the future.

Could this spontaneous force be Hoffman's immaterial network of conscious agents? Maybe this network is another form of intelligence, and through its will, it builds heterarchies, which our perceptual interface presents as subatomic particles, atoms, molecules, cells, and multicellular organisms. Perhaps the immaterial network isn't just a product of our simulation; maybe it is also the consciousness of that which is running the simulation. Perhaps it is the all-pervasive nonmaterial intelligence I perceived during my first meditation retreat.

Perhaps God isn't just playing dice; maybe he/she/it likes more extreme forms of gambling. Perhaps God loves fully immersive simulations. It seems like the game that is being played here is the organization of complex systems or the evolution towards ever more complex relationships. During each iteration, this order-seeking energy becomes more focused or is able to create more coherence in more complex systems.

This is a very unusual perspective because we are used to thinking that we live in a dead universe. But perhaps we are an expression of a larger living system which organizes information into ever-greater complexity and forms of identity. What we experience as love could be the multicellular expression of that which drives the evolution towards coherence, because if we follow the feeling of love, we build coherent social systems through harmonious relationships.

For physicists, spiritual statements like this are blasphemy, because they oppose the beliefs of materialism and connect our subjective experience to objective truths. If you ask mainstream physicists about the organization of subatomic particles, they might prefer to say that the strong force is responsible for holding neutrons and protons together, and the electromagnetic force keeps the electrons in orbit.

But if you look at these explanations carefully, you realize that they explain the rules of the game, not the player. The rules of a chess game will not make the pieces move. In other words, the laws of physics do not explain the subatomic coherence of an atom or the order within molecules. Why would these systems balance themselves? No one knows. There is a gap in this theory.

Brave scientists like Tom Campbell and Donald D. Hoffman are pushing the boundaries with their intellect, but perhaps mystical experience can also help mainstream science at this dead end.

The human being is a quantum system. If this quantum system is finely tuned through meditation, why shouldn't it be able to receive accurate information? Perhaps there is some truth to the experience of an all-pervasive network of love. Perhaps subatomic particles have free will and are balanced by the principle of love — the intention to create harmony together.

Conceiving the probability that drives the evolution of this universe as an intelligence that evolves towards love makes sense if we recognize that objective probability is free will. At the same time, we arrive at the essence of many religions and wisdom traditions. The history of religious belief is full of accounts from monks, priests, shamans and even laypeople describing experiences that showed them that the physical reality is an illusion and that beyond it is a loving oneness that expresses its will through the diverse life forms we find in this reality.

The idea that the physical universe is information which is organized by an evolving intelligence is in opposition to assumptions of materialism, but the insistence on materialism's fundamental truth isn't necessary for science. Science is a methodology, not an ontology.

Materialism has tried to avoid consciousness, but in the process, it only created more mysteries. What created the Big Bang? What is the force that creates order? Why are we conscious? Why is the speed of light fixed? Why does intention affect probability? Why does observation collapse the probability wave of a particle? Why do some particles decide their spin when observed? Why is there such a thing as the mystical experience? Why do healthy organisms create subatomic coherence? Why does the state of bliss result in brain coherence?

All of these questions remain mysteries in the materialistic doctrine but begin to make sense when we place consciousness at the center of reality. If matter is information which is organized by conscious agents, all these questions have logical answers.

If we compare these two world views, we realize that this expanded version of the simulation hypothesis isn't more mystical. It is more realistic.

French philosopher Rene Descartes is most well-known for his declaration *cogito ergo sum:* "I think, therefore I am." Descartes said the only thing that he can be sure of is that he is a thinking being. Everything else could have hidden variables which prevent one from seeing its true nature. But even if you are in a dream, you are still experiencing the dream. The experiencing element is still present, even if you don't know the nature or the origin of the experience. Therefore the fact that you are thinking is the only thing you can know with certainty.

Whether the external reality is really what we perceive it to be is unprovable. We are experiencing material objects, but we also experience objects in our dreams. Whether anything is what it seems to be is unprovable.

Materialism puts this fact aside and attempts to explain consciousness in terms of what's observable in the seemingly physical world. Besides the obvious uncertainty of the true nature of our experience, materialistic science has proposed an objective truth that doesn't take into account the only thing we can know with certainty: that we are thinking beings. If this is the only thing one can know with certainty, it would be a logical consequence to build a theory about the nature of reality on a solid foundation, not on an aspect of experience which cannot be proven with certainty. Therefore consciousness needs to be at the center of any theory about the nature of reality.

From this perspective, the simulation hypothesis and Donald D. Hoffman's ideas no longer seem completely outrageous. After we have recovered from the shattering of our cultural predisposition and beliefs, we must admit that this worldview seems to be more capable of explaining both our subjective experiences and the interactions we observe in the so-called "objective world."

CHAPTER 4:

Love and Fear

The evolution of fear and love in animals, fear vs love based organization of social systems

In Buddhism, the term "basic goodness" is used to refer to an innate quality of consciousness. Buddhists believe that when the mind is silenced through meditation, this intrinsic quality of goodness expresses itself.

Harvard graduate Dr. Richard Davidson conducted studies that he claimed proved that babies come into the world with innate goodness. These studies tracked the eye movements of infants with infrared cameras when they were presented with video clips of warmhearted encounters and video clips of hostile encounters. It was clear that the babies had more interest in the warm-hearted encounters. Davidson used this evidence to argue that there is an innate tendency in a baby's consciousness to prefer kindness and love.

Buddhists believe that this basic goodness is a fundamental element of consciousness. If this is true, why is there so much fear and evil in this world?

Whenever individual conscious agents make choices in relation to one another, two possible consciousness modalities can be expressed: the consciousness modality of love and the consciousness modality of fear. Love-based choices create organization in complex systems that are based on collaboration, while fear consciousness creates organization through domination. The modality of love consciousness is based on choices that include the will of others and build relationships willingly. The modality of fear consciousness diverts and overrides the will of others and builds relationships against the will of others.

Animals have evolved fear-based behavior patterns to take resources from each other, protect themselves from predators, and outdo rivals. Animals that did not have fear could not effectively identify danger and therefore had an evolutionary disadvantage when they had to face a predator. A good example is the extinction of the dodo bird. The dodo was a flightless bird, native to the island of Mauritius, near Madagascar. It became extinct in 1662 because Dutch sailors took advantage of the bird's lack of fear. There were no predators on the island, so the bird did not need to evolve fear-based behavior mechanisms. When Europeans came to the island, this lack of fear-based behavior became the bird's downfall. The story goes that the sailors fatally clubbed the tame birds until none were left.

It is rare to find animals that don't have fear, since most animals had to evolve defensive strategies to keep their species alive. Gazelles run away from lions, mice run away from cats, and so on. But the phrase "fear consciousness" doesn't only refer to defensive behavior. Fear consciousness also includes aggressive behavior or any other behavior which diverts the will of another being. The reason I call it fear consciousness is because the control chains of this type of consciousness build on the emotion of fear. If you are afraid of something that threatens you, you empower the will of whatever threatens you.

As far as I know, all animals that are alive today have evolved fear. But many animals have also developed love-based instincts, since love, the will to benefit others, allows more information to be passed down from one generation to the next.

For example, all mammals care for their offspring and teach them how to gather food or how to hunt. Animals create organization within their social environment by relating to one another based on the principles of sharing and caring. For example, chimpanzees groom each other, and bonobos solve conflicts by having sex. Animals who treat each other with benevolence form strong bonds, and through those bonds, the pack or tribe becomes more organized and coherent. At the same time, animals still have all these fear-based behavior mechanisms that are also necessary for the survival of the species.

The difference between these two behavior mechanisms or consciousness modalities is the source of the struggle between good and evil. Love-based consciousness produces benevolence, compassion, patience, collaboration, and all other virtuous principles and attitudes, while fear consciousness produces violence, aggression, anger, manipulation, and competition, which are behavior patterns based on protecting the self. Manipulation might sound like it does not belong in this category because it is usually associated with careful planning and not with animal instincts. But the pattern I am trying to point out is based on the relationship between free will units, not the specific manifestations of that relationship. Violence, aggression, anger, manipulation, and competition all are based on dominating another being's will. Love is about integration. That is the fundamental difference between the consciousness modality of fear and the consciousness modality of love.

Fear consciousness is based on overriding, diverting, distracting or outperforming the will of another free will unit, while love consciousness is based on including, supporting, and integrating the intention of another unit. This definition of fear consciousness might sound counterintuitive because fear is usually associated with running away, but running away from a predator is also driven by the intention to override the predator's will.

Fear and love are two different organizational principles that become stratified moral systems as relationships gain complexity. For example, an animal attacks another animal, while a country plans how to invade another country. An animal's attack is spontaneous and primitive, while a country's military action is planned and sophisticated. Although fear is a basic survival drive, it has woven itself into the complex power structure of today's society.

In today's United States, the military is a large organization that specializes in diverting and overriding the will of other nations. But the military is just one tentacle of a giant fear-based network. The media, the economy, the school system, the medical industry and the legal system all have a high level of fear consciousness and perpetuate this type of consciousness through content, practices, and laws. The media features stories that are fear-inducing, the economy is based on scarcity, the school system uses competition, the medical industry is heavily influenced by the selfish interest of big drug companies, and the prison industrial complex is driven by private profits.

Selfish desires are also part of fear consciousness. For example, greed is created by the fear of not getting enough, and the desire for power comes from the fear of being helpless or the fear of being weak.

Complex systems are most efficiently organized by love consciousness, but fear consciousness can also organize complex systems by creating layers of fear. We don't see a pure form of this type of organization in today's society, because people are also driven by love consciousness and the will to benefit others. However, there are expressions of fear-based organization in today's society. People are driven by the fear of not being good enough, not looking good enough, not having enough, and so forth.

Whenever people do something because they are afraid of something else, we find fear-based actions that create fear-based systems. This type of system is much less efficient than a love-based system. The leaders of a fear-based system need central control and more laws and more law enforcement to keep people organized, since all the organization occurred against their will. If people don't want to work together or don't want to benefit each other, contracts need to be signed, court dates need to be scheduled, and laws need to be enforced.

On the other hand, if people work together because they love it, you don't need an external force to create order — the will to create harmony together becomes the organizing principle. In a love-based system, you need fewer rules and less law enforcement because there is less conflict. However, it is easier to achieve a fear-based organization in a primitive civilization because you don't need mutual agreements to create a fear-based society. If the oppressor chooses to oppress and has enough power to do so, a fear-based society is the result.

A love-based society needs mutual agreements. All countries and all people would have to come together with the collective interest of helping each other.

Whether a love-based society will be created by humans or another species depends on what all of us choose.

It seems that evolution favors harmony and coherence in the long run. The unlikely events eventually occur and perpetuate themselves through their coherence. It seems likely that intelligent multicellular beings would eventually build a global system that takes every being's will into account and creates a coherent whole. Just like it was only a question of time until the unlikely events occurred that made cells collaborate.

The patterns I just discussed are based on principles that are true whenever you have free will in a reality with a limited rule set, such as space and time. Evolution is driven by free will units that interact and compete for resources. In competition, the structures of fear and love always emerge. I can say "always" because it is a logical consequence that is not dependent on the details or the content of the system — it is true on the level of relationships and will. So even highly advanced aliens will either build fear- or love-based systems, or complex mixtures of both, as long as they function within a limited rule set. If they have free will and relate to one another, they need to decide how to deal with the intention of another being: either integrate it or divert it.

It is fair to assume that old civilizations end up building love-based systems because it is almost impossible to create stable fear-based societies. Fear-based systems have a very high probability of collapse because they are brought together by an inverted will. If people work together but are focused on selfish gain, then the system is built on a contradiction, and revolutions and wars constantly threaten its organization.

Free will provides the option of relating to others through fear or love. Evolution rewards both in different ways. Fear has short-term benefits in terms of survival of an individual; love has long-term benefits in terms of the survival of a group or a civilization.

The Buddhist idea that love consciousness, or basic goodness, is a fundamental aspect of awareness can be integrated into evolutionary psychology because fear consciousness can be seen as the product of love consciousness when it evolves within the illusion of separation and under the constraints of a limited rule set. In that light, good creates evil when it operates blindly. Even if all people and all beings are driven by love consciousness, they create fear consciousness by protecting their illusory identity from the scarcity that is simulated in this "physical" reality.

From the Buddhist perspective, this isn't accidental. Through the identification with the illusion, love consciousness explores new relationships through which it learns and grows. Buddhists believe that this is the purpose for our physical life cycles. We'll explore this idea in greater detail later on in Chapter 24.

CHAPTER 5:

Fields of Dreams

The idea of a soul and the experiences of an identity beyond the physical body, Rupert Sheldrakes morphogenetic fields, the chimpanzee genome project and the genes of a rice plant

The exact meaning of the word "soul" varies somewhat by culture, but in general it refers to an eternal and essential part of a human being. Religions popularized the idea of a soul, but later that concept was rejected by scientific materialism and positivism. However, science hasn't dismissed the idea of a soul based on empirical evidence; rather, the scientific community expected that the entire human organism would eventually be explained in purely materialistic terms.

To put it mildly, that hasn't happened quite yet. So far the nature of consciousness remains a mystery that science hasn't been able to explain, and the idea of a soul hasn't been proven or disproven.

At first glance, proving the existence of the soul seems impossible, and most scientists would rather not venture into such mystical territory. The reason I believe that this exploration is worth the effort is due to the personal experience I briefly mentioned in Chapter 1.

I always believed that I only existed in my brain, but as I sat for days in silence, I had the overwhelming experience of recognizing my eternal nature. Days of silent meditation brought me into a state of consciousness in which my human existence, my everyday identity, was much further away from me than my essence, which came forward and demanded to be recognized as an eternal being.

Although this experience didn't leave me with any empirical evidence, it seemed to reach a part of myself that functions beyond rational understanding. During meditation, one empties one's mind until there is a complete absence of rational processes. It was in this altered state of consciousness that I first had an encounter with a part of me that appeared to transcend everything I had learned during my life. I had experienced something that was unprovable, yet it felt more real than any truth I could have obtained through rational contemplation or empirical evidence. Either this experience was a convincing illusion, or meditation had given me a glimpse of a mystical truth that lies beyond the understanding of today's science.

Any radio technician knows you need a clean signal to transmit information. Perhaps one way of looking at it is that the brain is the "receiver" of the soul, and brain coherence allows us to receive a cleaner signal. Maybe meditation is a tool to gain insights from a source that is usually obscured by the psychological noise of everyday life.

Many long-term meditators have described experiences similar to the one I had, and sometimes these experiences occur independent of meditation. Neuroscientist Andrew Newton's book *How Enlightenment Changes Your Brain* analyzes dozens of case studies regarding "enlightenment experiences." These experiences could be described as nonverbal or non-rational epiphanies with life-changing consequences.

Newton argues that the significance of these experiences exists in the experience itself. He believes that it isn't important whether these experiences originate beyond our physical reality or whether they are a function of the brain. This is probably true, since believing one theory or another doesn't reproduce the experience. However, I think it is important for the advancement of physics, psychology, biology, and theology to logically explore the possibility that our identities go beyond the human brain and the personalities we create during our physical life.

I believe that the simulation hypothesis is the most likely candidate for bridging the gap between mystical experiences and empirical evidence and providing a comprehensive framework to subsume the work that humanity has done so far in the different areas of scientific and mystical explorations.

In earlier chapters, I talked about the universe as an evolving information system driven towards ever-higher complexity by an interconnected web of awareness which seeks to express harmony. On an atomic and subatomic level, this harmony is represented by minimum energy states, and on a cellular and multicellular level, this harmony is expressed by collaborative systems. This hypothesis further proposes that there is a nonphysical element which opposes entropy and organizes the information we perceive as matter.

From one perspective this energy is a continuum or a whole, but from another perspective, it has within itself different units. In the same way, a river is one unit, but within the river are waterfalls and standing waves. We can recognize the river as a whole and also the waves and waterfalls as individual units. Your individual identity is like a standing wave in a web of interconnected quantum fluctuations.

Physicist David Bohm defines all things, from particle to observer, as "semi-autonomous quasi-local features" — meaning that they are only partially conceivable as separate from the whole, just like we can never entirely separate a standing wave from a river but can still recognize it as a unit.

Perhaps we can view the idea of a soul from this same perspective. Perhaps the soul is an information pattern in the nonphysical web of conscious agents.

British biologist Rupert Sheldrake approaches the same hypothesis with different terminology. He conceives the nonphysical energy of the universe as non-local fields of information. By non-local, he means that such fields are not constrained by time or space. Sheldrake calls these information fields "morphogenetic fields" and proposes that any form-based structure in the universe is linked to a nonphysical information field which holds the records of all previous manifestations.

Under Sheldrake's hypothesis, a silicon crystal has a morphogenetic field that contains the information necessary for arranging the atoms and molecules in that particular pattern. Likewise, a certain type of tree has a morphogenetic field which contains all the information created by all trees of that species.

Unsurprisingly, Sheldrake's ideas are not particularly popular among traditional materialistic scientists. But after reading his dense books filled with extensive experimental evidence, it seems that morphogenetic fields are a better explanation for the development of form in organisms and non-organic structures.

For example, in his book *A New Science of Life*, Sheldrake writes about *turanose*, an artificially created sugar that changed the parameters of its crystallization process simultaneously at laboratories all over the world. There are several other examples of the interconnected evolution of crystals. Xylitol was a liquid sweetener in 1891 until it became a crystal all over the world in 1942. Its melting point was originally 61 degrees Celsius and then rose a few years later to 94 degrees.[3]

The pharmaceutical industry often has to deal with the unstable nature of newly invented crystals. For example, Abbott Laboratories lost an estimated 250 million dollars when an ingredient of its AIDS drug, Ritonavir, suddenly and unexpectedly shifted its crystalline structure even though the production process remained the same, forcing the company to pull the drug off the market.

These types of findings have baffled many scientists. Sheldrake believes that newly invented crystals, or crystals that are suddenly mass-produced, can abruptly change their shapes or the temperature at which they crystalize because their morphogenetic fields are less stable. He believes that any repetitive activity in the universe creates an information field that is in a constant data exchange with the form the activity created. In this understanding, crystals add and draw from a morphogenetic field and that field becomes more stable through the formation of each crystal. This would explain why newly invented crystals suddenly change their properties and crystals that have existed for billions of years always crystallize in the same way.

Quantum physics has long shown that particles can be entangled and share instantaneous information independent of time and space. It isn't a far stretch to propose that an organism might use this capacity of the universe for the purpose of morphogenesis.

Morphogenesis is the process of the growth of forms. Surprisingly, no one can explain how organisms take on specific forms. Biologists assume that DNA is responsible for the shape of an organism, but

---

[3] Holden and Singer (1961), pp. 80-1.

this connection has never been explicitly proven.[4] We only know that DNA is responsible for protein production, but that doesn't determine the specific forms. For example, our feet and hands contain the same proteins but have a different form.

Experiments with sea urchins have shown the process of morphogenesis to be even more mysterious. In Hans Driesch's classic sea urchin experiments, one half of a young sea urchin embryo was killed. The second half adjusted and formed a small, but complete sea urchin. The reverse was true when two embryos were fused. The system adjusted and built one giant sea urchin.

Materialistic biology attempts to conceive of animals and plants as complex machines. However, try to imagine a machine that builds itself, is then cut in half, and then proceeds to adjust to the damage by creating a smaller version of itself. This machine would have to have communication between all of its parts and a way to make collective decisions to work towards a collective goal.

Sheldrake proposes that this communication within organisms happens through an information field which all cells in the organism share. He says that DNA is only responsible for manufacturing proteins, but the form of an organism is constructed via an immaterial information network — the morphogenetic field.

Though scientists firmly believed that DNA is solely responsible for the nature of an organism for the first half of the 20th century, the rise of gene sequencing has caused some to revisit that belief.

The findings of the Chimpanzee Genome Project, published in 2005, were supposed to show why a chimpanzee is so different from a human being. However, the opposite occurred; after the team finished

---

4 Rupert Shedrake, *A New Science of Life,* J P Tarcher, Inc.

sequencing the chimpanzee genome, the director of the project said, "We cannot see in this why we are so different from chimpanzees."[5]

If genes are responsible for the complexity of an organism, it would follow that the more complex an organism is, the more genes it has. This isn't the case, though. A human has about 25,000 genes, a sea urchin has about 26,000, and some plants have even more. Rice has about 38,000 thousand genes!

So what is it then that makes the human so different from the chimpanzee, or one organism more complex than another?

A good analogy to make sense of this is the relationship between building material and an architectural plan. You can build a very complex house with just one type of cinder block. The complexity of a house doesn't necessarily come from the different building materials one uses. The architect's design choices have a much greater effect on the house's appearance. The architect's design is just like a morphogenetic field; it is information. We can also call that design a "soul" if we see the soul as an information network that organizes and orchestrates processes in physical beings.

In the following chapter, we'll look at out-of-body experiences to examine this hypothesis on a practical level. If we truly have souls, we should be able to have experiences that aren't confined by our physical reality.

---

[5] Olsen and Varki 2004

CHAPTER 6:

Body, Soul, and Awareness

The international out-of-body survey, Robert Monroe's out-of-body research lab

Topics like out-of-body experiences are often enough to make a traditional scientist close a book. But again, science isn't an ontology; it is a methodology. So any phenomenon can be scientifically examined as long as the investigation process is logical and rigorous.

Radio broadcasting executive Robert Monroe's 1971 book *Journeys Out of the Body* popularized the term "out-of-body experience" to describe the experience of a separation between body and awareness. According to Monroe, he spontaneously had the experience of leaving his body one night in the late 1950s. He was fully awake and was floating above his sleeping body. For a year he kept having experiences like this, which frightened him tremendously, but he eventually gained the courage to explore these novel states of consciousness.

Over the following decades he wrote three books on the subject, founded a research facility named Monroe Industries, and produced case studies that made many people question the nature of reality.

In his research facility, Monroe set up heated waterbeds with headphones next to them. He would ask his test subjects to lie down and listen to soundtracks that he claimed helped people leave their bodies.

Some of his test subjects later described the experience of traveling through otherworldly landscapes and encountering different forms of intelligence.

I read the transcripts of many of these case studies and interviewed Tom Campbell about his time at the Monroe Institute. Although the phenomena they examined seemed outright crazy, they approached the subject in a very scientific way.

In the 1970s Robert Monroe was trying to figure out whether out-of-body experiences are objective or subjective phenomena. Put another way, he was wondering if the experiences are imagined or whether two people can experience the same reality during an OBE. This is a critical question, because if two different people can independently describe a shared out-of-body experience, it would mean that an individual's imagination does not produce out-of-body experiences. Monroe had a team of people who claimed to be able to leave their bodies. He set up a clever experiment.

After a year of experimenting with OBEs, some of Monroe's test subjects — or "explorers," as he referred to them — had developed the ability to communicate via their physical bodies while having the experience of traveling through another realm or dimension.

In one of Monroe's experiments, two travelers were lying in separate rooms and attempted to leave their bodies and meet each other. Through headsets and microphones, they could communicate with Monroe, but could not hear or see each other. Once the subjects reported having left their bodies, Monroe periodically asked them what they were experiencing and recorded each conversation on a separate tape. The explorers reported that they had met up successfully and gave detailed descriptions of their surroundings and the non-physical beings they encountered. After an hour and a half, Monroe asked them to come back to their bodies and meet him in the control room. One of the two travelers was Tom Campbell.

Campbell said that his approach to the whole project could be called "open-minded skepticism." Although Campbell had gained the ability to leave his body voluntarily and perceived himself to be traveling through other realms, he didn't know if what he was experiencing was a subjective fabrication of his mind or a reality others could experience as well.

After Monroe had asked him to come back to his body, Campbell recalls the lights being painfully bright. Monroe started playing both tapes simultaneously, and the two explorers' experiences matched perfectly. Campbell said that after this experience he kept repeating to himself, "THIS IS ACTUALLY REAL!"

William Buhlman is another scientist who became obsessed with out-of-body experiences. While studying psychology, he started practicing out-of-body techniques, eventually claiming that he was having OBEs several times a week.

Buhlman wrote about OBEs in three books. In the first, *Adventures Beyond the Body*, he wrote about how he learned to leave his body and shared techniques that helped him. For his second book, *Secret of the Soul*, he conducted an international out-of-body experience survey. 16,000 people from over 42 countries participated, and Buhlman published some of the testimonies.

To get a sense of what out-of-body experiences are like, let's look at a few testimonies from the survey:

> *I am not a believer in this kind of stuff, but I decided to try one of the techniques from [Adventures Beyond the Body] for about a month. After about three weeks, weird things started to happen. A loud engine sound woke me up*

several times during the night. The next day I was scared awake by sounds, and I couldn't move my body for several minutes. I hate to admit it, but this was really scary to me... Anyway, I continued the techniques every night for two more weeks and then "it" really started to happen. I woke up completely paralyzed, but this time I stayed calm and thought about the door. Bang! I was suddenly standing by the steel door, and I could see through it. The door and the walls were like layers of fog. I touched the fog and then stepped through it. I was suddenly in a new place, a green field. I was so shocked by the change that I seemed to be sucked back into my body. I opened my eyes and felt paralyzed for a few seconds. Then it all hit me like a bolt of lightning: It was completely real. I lay there in total amazement: 'This is bigger than anything I can imagine; how can I tell people without sounding like a nutcase? This changes everything I ever thought I knew or thought was true. I need to know more, see more.' —T. R.[6]

Several months ago I did an out-of-body technique every night as I fell asleep. I was very focused on finding answers and having a spiritual experience. One night I fell asleep normally and was shaken awake by what I thought was a train running through my bedroom. I tried to move but couldn't. I then thought about knowing God. I instantly shot through an incredible series, or layers, of colors and was suddenly floating in an endless ocean of white light.

---

[6] William Buhlman, Secret of the Soul, (HarperOne, December 6, 2011)

I was connected to everything and everyone. —Eric L., Detroit, Michigan[7]

When I am having my most profound out-of-body experiences, I move through a rainbow and my entire being is completely surrounded by an endless sea of liquid color filled with loving light. I feel as if I am the light, but yet somehow I am still an individual. An incredible feeling of peace and love surrounds me. I'm part of the universe. All knowledge is present.—Alan D., Palm Beach Florida[8]

Materialistic scientists would argue that these types of experiences are the result of some not-yet-understood process in the brain, while more integrative scientists like Rupert Sheldrake and Tom Campbell would say that these experiences occur when consciousness separates from the data stream of the body and merges with a larger field of awareness. Perhaps the liquid sea of white light is the direct experience of Donald D. Hoffman's immaterial network of conscious agents, or Tom Campbell's "larger consciousness system."

We could view these type of OBEs as the experiential counterpart to the hypothesis we discussed earlier. The idea that consciousness is an all-pervasive nonphysical energy is not only a theory but also an experiential truth for many people.

In an interview at the London Real, Rupert Sheldrake tells a story about how people in Cambridge laughed at him when he first suggested that consciousness creates nonlocal information fields and guides the process of morphogenesis. This idea was so strange to his colleges that no one took him seriously.

---

[7] Op. cit.

[8] Op. cit.

Sheldrake later moved to India for a job, and said when he told his friends there about his hypothesis, people would also laugh at him — but this time for the opposite reason. They would say, "This is nothing new." The idea that everything is part of an interconnected field of awareness is a common belief in India — the second-most populous country in the world. Many citizens practice meditation and deliberately induce mystical experiences. However, this is not the case in Western culture, which does not yet have a framework to make sense of mystical experiences.

Materialistic science has been so successful in explaining physical phenomena that materialism has become a type of religion. There are so many details to learn about every aspect of how matter behaves that we forget that materialism is an unproven hypothesis that cannot explain why we are conscious, let alone what consciousness is.

I am not saying we should just believe every crazy theory. We have to be skeptical of believers of any kind. The increasing popularity of Eastern religions and philosophies in the West has created many pseudo-scientists who throw around terminology but don't work methodically. They are an insult to science. We should not assume that an experience represents reality, or that an ancient belief system is true. Even if we have incredibly positive experiences, such as traveling through infinite oceans of loving light, we still have to ask: how can we be sure that these experiences aren't just fabrications of our brain?

CHAPTER 7:

Field Tests (I)

My personal out-of-body experiences

While doing research on out-of-body experiences, I became interested in exploring the topic on a personal level. Most of the techniques for inducing out-of-body experiences are based on visualizations and affirmations. When you're falling asleep, you intentionally visualize something, or you tell yourself over and over again, "I will leave my body tonight!" I tried the visualization techniques; for the first couple of weeks nothing happened, but then things got very strange.

I fell asleep and dreamed that I was sitting in a circle with a dozen other people. In the middle was a small human-like creature that danced around, waved its hands, and made sounds that I could not understand. The creature approached every person in the circle and appeared to be putting everyone into a trance. In the dream, I had a rebellious attitude toward this ceremony, or whatever this strange gathering was, and felt certain whatever the creature was doing to the other people would not work on me.

When the creature got to me, it put one hand on each side of my head and smacked both of my ears. I woke up startled — but I was not in my body. I was floating in complete darkness. An intense current of crackling electricity was shaking my entire being, and I felt the hands of the creature holding my head. "Calm down, calm down, I am getting you out of your body," it said with a raspy voice.

I could not relax. It was dark and uncomfortable. Instead, I woke up in my body, somewhat disappointed that I could not relax.

I had read about many case studies in which people described encountering nonphysical beings who tried to help them get out of their bodies. For some reason the idea didn't freak me out. I kept practicing the visualizations and kept telling myself, "I will learn how to leave my body."

A couple of nights later I was dreaming that a large snake was swimming through a pond. Suddenly, I became the snake. As I swam through the pond, my awareness became brighter until I was fully awake. I was submerged in murky green water. It was so real that I started to panic. I felt the need to breathe, but I did not want to inhale any water. Then I realized that I was not in my body; the desperate need to breathe disappeared, but I also started to become aware of my own physical body, which was asleep.

Once again, I was in the uncomfortable in-between state. I could hear and feel something like an electric current and everything became black again. I could feel a being holding my feet and pressing its thumbs into my soles. It seemed like the being was regulating my energy field by sending a strong current into my feet.

I could not relax, and I landed disappointed back in my body. When I woke up, I could feel a tingling sensation on my feet, right where the being had been holding them.

I kept practicing, determined to have a proper OBE.

A few months later a similar thing happened again. I was sleeping at a friend's place and as I was falling asleep, I practiced my visualization exercises. I visualized the mirror in the bathroom and imagined touching the wooden frame and looking at myself. I tried to make the image as real as possible. I began to dream, but I knew I was dreaming. At some point, the dream became very real, and I forgot that I was dreaming.

I found myself in Tamera, a peace research center in Portugal. I had a little room there with a bed, and I laid down to practice my out-of-body technique. Now I was in a dream trying to leave my body. Suddenly the entire image froze. I felt a presence and the word "hello" flashed in front of my eyes. I felt hands gently grabbing my feet and thumbs pressing into my soles. The energy that seemed to come out of these thumbs was shaking my entire body (which I thought was in Portugal). The energy was so intense that I wanted to open my mouth and scream, but I knew I had to get through these sensations if I wanted to get out of my body.

Suddenly, I was in a parking lot. I could no longer perceive the being, and I was fully aware that I wasn't in my body. The first thing I did was fly straight up, just by imagining it. When I was pretty high up, I flew towards the ground as fast as I could, and right before I hit the ground, I made a sharp curve and flew back up. I was flying in a loop as fast as I could and eventually I got very dizzy.

I landed in my first dream, in my bed in Portugal, and I could feel the being again as it massaged my feet. Then it gently placed my left foot next to my right knee. I heard a dog barking, and I woke up back in my body with my left foot next to my right knee.

All day long my feet tingled, and I could still feel the points where the being's thumbs had touched my soles. It was strange to feel the physical effects of my preliminary OBEs, but I kept practicing the technique.

A few weeks later I was still very determined to have an OBE. Before I went to bed, I stood in front of the mirror in the bathroom, and for several minutes I tried to remember exactly what I was seeing. I went to bed and focused on details I had paid attention to: the shadow my nose was casting on my lips, the light reflecting on my hair, and the doorframe reflecting in the mirror.

At some point, I fell asleep and then became fully conscious, but I was no longer in my body; I was in the bathroom. I started to walk back into the room I was sleeping in and noticed that I didn't need to move my legs. I would think about moving and immediately start to glide forward. I came up next to my sleeping body and became overly excited. I had finally managed to leave my body! I looked ahead and saw that the bedroom was opening up into a vast space of patterns and otherworldly designs. This startled me so much that I became aware of my beating heart. I was back in my body.

I had another OBE experience when Moon Hooch played at the Joshua Tree music festival in 2015. A couple of hours before the show I was lying in our hotel room. We had taken an overnight flight from Indianapolis, where we had played the night before, and I was exhausted but too excited to sleep. We were about to play in front of a couple of thousand people, and there was supposed to be a huge laser show.

I laid down and started visualizing myself walking around. Suddenly I was in a lobby, and a security guard in a striped shirt was standing next to me. I asked him if he could hear me.

"Yes," he replied, "but I know you are not real."

"What do you mean?" I asked.

"I know you are just in my head."

"No, I have a body, I put it to sleep in a hotel in Palm Springs. I swear I am real," I insisted.

"No, you are not. I am going crazy. I am hearing voices." The whole time he was talking to me without opening his mouth. I wanted to assure him that he was not going crazy so I asked him if there was anything only he could know and that I would try to find him when I woke up to confirm my existence.

"Yes," he said. "I used to have a passion for martial arts, but that was a long time ago, now I just sit here and do nothing." When I woke up, it was time to leave.

I still don't know if these experiences were objectively true or partially imagined, but they gave me a different perspective on what could be happening. Tom Campbell's words now took on a new meaning: "I am working with nonphysical beings on shifting the probable outcomes in different realities."

I found the idea that one could leave one's body and influence others to be both spooky and interesting. Yet it occurred more and more often that I found myself being somewhere between the dream world and the "real world," having educative discussions with all sorts of beings.

Once, I met a monk who told me that what he wanted to teach me cannot be put into words. He looked at me and transmitted some type of energy that he said I should use to improve my interpersonal relationships.

Other times, I found myself in the teaching position. I found myself passing on the intellectual frameworks of this book to humanoid beings that had lost access to their spiritual connection and appeared to be in the need of a rational framework to find their own essence. During one of those experiences, I looked at myself in a mirror and was shocked to see that I seemed to have a very long light pink skull with huge gray spikes on the top.

Yet another time, I was going to sleep and became aware of a light beam coming from the ceiling. The light beam was hollow and I traveled into it. I felt a pulling sensation in my heart and gradually the outlines of a high tech train station appeared. All sorts of alien-looking beings were getting on and off the train. I was a nonphysical observer in that reality and no one noticed me. All these experiences were definitely shifting the probable outcomes of my life, because they changed my worldview and my values.

When I was dating my ex-girlfriend, Julia, we often talked about these kinds of things. She too had very strange otherworldly experiences somewhere between dreams and out-of-body experiences. But we both felt that these experiences might be pointing towards the existence of a larger reality. This feeling became especially pronounced when we tried to visit each other by leaving our bodies.

I was in the Pacific Northwest on tour, staying in a house on the Columbia River near Seattle. I was taking a nap while she was meditating in her room in New York City and tried to meet me. Suddenly she saw a flash of light and found herself traveling through several layers of colored light. Then she saw me taking a nap. She embraced me and gave me a hug. At that moment her face appeared in my dream. She was radiating love and kindness, and it felt so real that it woke me up. She called me not much later and asked me if I saw her. We were both pretty freaked out. It is one thing to read about case studies that talk about these kinds of things and quite another thing to experience them directly.

CHAPTER 8:

On the Threshold

Near-death experiences

Near-death experiences, or NDEs, are very similar to out-of-body experiences, except the person's body is clinically dead during the experience. Clinical death means that breathing has stopped and there is no measurable activity in the brain or heart.

I came across NDE research while I was studying OBEs. As I read through the comments on an article about OBEs, I encountered an interesting discussion. One person said that he had an out-of-body experience and that he was not sure if it was a hallucination or an extrasensory experience, and he posted a few links to testimonies of near-death experiences suggesting that when the body dies people also have out-of-body experiences.

One of the videos he posted featured a story of a patient surviving cardiac arrest. After the patient was resuscitated in the hospital, she said that while she was unconscious she floated out of her body and saw a blue tennis shoe on a ledge on the third floor. She talked to the nurse about it and asked her to check if there was a blue tennis shoe on the third floor. The nurse went upstairs and indeed found a blue tennis shoe on the ledge on the third floor. There are many stories like this one, and there isn't an explanation for how a clinically dead person can obtain this kind of information.

A near-death experience is typically associated with the subject reaching clinical death and being resuscitated several minutes afterward. Thousands of people all over the world claim to have experienced NDEs. Since 1975, over 3,000 detailed scientific studies have dealt with the phenomena, and there have been millions of personal testimonies.

Bruce Greyson, Professor Emeritus of Psychiatry and Neurobehavioral Sciences at the University of Virginia, established the Greyson Scale in 1983 by comparing and categorizing a large number of testimonies.[9] Researchers use the Greyson Scale to determine if a specific subject had an NDE and how "deep" it was.

Greyson reduced 80 characteristics of the NDE to sixteen elements and placed them into four groups: cognitive, affective, paranormal, and transcendental. Losing one's sense of time, having accelerated thoughts, life review, and the experience of universal knowledge are cognitive effects of the NDE.[10] Feelings of peace, joy, cosmic unity, and the perception of a brilliant and non-blinding light are components of the NDE that he classified as affective. Paranormal refers to all phenomena that can't be explained by the materialistic worldview. The paranormal component may include accurate auditory and visual perceptions, the experience of remote events, premonitions and prophetic visions, and an out-of-body experience. The transcendental element involves the experience of unearthly realms, meeting or sensing the presence of mystical beings, seeing and communicating

---

9 Bruce Greyson, *the handbook of near-death experiences,* (Praeger June 22, 2009)

10 Dr Pim van Lommel, *Consciousness Beyond Life*, (HarperOne, June 8th 2010)

with deceased persons or religious figures, and reaching a border that cannot be crossed if the person intends to return to the body.

There are hundreds of case studies of NDEs and hypnosis-induced past death memories that describe tunnels and lights. Sometimes the tunnel opens next to them, other times they float high above the earth before the tunnel appears.

One of the most famous cases of an NDE was featured in a BBC documentary *The Day I Died* and published by Dr. Pim van Lommel in his 2010 book *Consciousness Beyond Life*. This NDE demonstrates many of the components covered by the Greyson Scale.

Pam was a working mother who had carved out a career as a singer and songwriter. In 1991 she became ill, and a CAT scan revealed a giant aneurysm at the bottom of her brain. She was told that her chances of survival were very slim. She contacted neurosurgeon Dr. Robert Spetzler at the Barrow Neurological Institute in Phoenix, Arizona. Despite all odds, he decided to perform surgery on her. Here is what Dr. Spetzler said: "What we're looking at is the aneurysm that she had, which is at the very base of the brain. This is the balloon that can burst and cause an incredible catastrophe in the patient's brain. This is why it was so difficult in this particular case.... What we want to do is we want to bring that brain to a halt. We don't just want the brain to be asleep. We want the metabolic activity of the brain to stop. Every measurable output that the body puts out really disappears completely so that you have no measurable neuronal activity whatsoever. Prior to the operation starting, a lot of activity goes on. The patient is put to sleep, the eyes are taped shut, and there are little clicking devices put in each ear in order to monitor the brain. The patient is then completely covered; the only thing that's really exposed is the area of the head where we work."

Cardiologist Michael Sadom said, "During standstill, Pam's brain was found dead by all three clinical tests—her electroencephalogram was silent, her brain-stem response was absent, and no blood flowed through her brain.... Her eyes were lubricated to prevent drying and then taped shut. Additionally, she was under deep general anesthesia."

Here is what Pam experienced: "I don't remember an operating room. I don't remember seeing Doctor Spetzler at all. I was with a fellow; one of his fellows was with me at that time. After that...nothing. Absolutely nothing. Until the sound...and the sound was...unpleasant. It was guttural. It was reminiscent of being in a dentist's office. And I remember the top of my head tingling, and I just sort of popped out of the top of my head. The further out of my body I got, the more clear the tone became. I remember seeing several things in the operating room when I was looking down. I was the most aware that I've ever been in my entire life. And I was then looking down at my body, and I knew that it was my body. But I didn't care. I thought the way they had my head shaved was very peculiar. I expected them to take all of the hair, but they didn't.

I was metaphorically sitting on Dr. Spetzler's shoulder. It wasn't like normal vision. It was brighter and more focused and clearer than normal vision. There was so much in the operating room that I didn't recognize, and so many people. I remember the instrument in his hand; it looked like the handle of my electric toothbrush. I had assumed that they were going to open the skull with a saw. I had heard the term saw, but what I saw looked a lot more like a drill than a saw. It even had little bits that were kept in this case that looked like the case that my father stored his socket wrenches in when I was a child. I saw the grip of the saw, but I didn't see them use it on my head, but I think I heard it being used on something. It was humming at a relatively high pitch. I remember the heart-lung machine. I didn't like the respirator.... I remember a lot of tools and instruments that I did not readily recognize. And I distinctly remember a female voice saying: 'We have a problem. Her arteries are too small.' And then a male voice: 'Try the other side.'

It seemed to come from further down on the table. I do remember wondering, 'what are they doing there [laughs] because this is brain surgery!' What had happened was that they accessed the femoral arteries in order to drain the blood, and I didn't understand that.... I felt a "presence." I sort of turned around to look at it. And that's when I saw the very tiny pinpoint of light. And the light started to pull me, but not against my will. I was going of my own accord because I wanted to go. And there was a physical sensation to the point where... and I know how that must sound... nonetheless, it's true. There was a physical sensation, rather like going over a hill real fast. It was like The Wizard of Oz—being taken up in a tornado vortex, only you're not spinning around. The feeling was like going up in an elevator real fast. It was like a tunnel, but it wasn't a tunnel. And I went toward the light. The closer I got to the light, I began to discern different figures, different people, and I distinctly heard my grandmother calling me. She has a very distinct voice. But I didn't hear her call me with my ears.... It was a clearer hearing than with my ears.... And I immediately went to her. The light was incredibly bright, like sitting in the middle of a lightbulb. I noticed that as I began to discern different figures in the light—and they were all covered with light, they were light, and had light permeating all around them—they began to form shapes I could recognize and understand. And I saw many, many people I knew and many, many I didn't know, but I knew that I was somehow, and in some way connected to them. And it felt...great! Everyone I saw, looking back on it, fit perfectly into my understanding of what that person looked like at their best during their lives. I recognized a lot of people. And one of them was my grandmother. And I saw my Uncle Gene, who passed away when he was only thirty-nine years old. He taught me a lot; he taught me to play my first guitar. So was my great-great-aunt Maggie. On Papa's side of the family, my grandfather was there.... They were specifically taking care of me, looking after me. They wouldn't permit me to go further.... It was communicated to me—that's the best way I know how to say it because they didn't speak like I'm speaking—that if I went all the way into the light something would happen to me physically. They would be unable to put me back into the body, like I had gone too far and they couldn't reconnect. So they wouldn't let me go anywhere or do anything. I wanted to go into the light, but I also wanted to come back. I had children to rear. It was like watching a movie on fast-forward on your VCR: You get the general idea, but the individual freeze-frames aren't slow enough to get detail.... I asked if God was the light, and the answer was: 'No, God is not the light, the light is what happens when God breathes.' And I distinctly remember thinking: 'I'm standing in the breath of God....'

At some point, I was reminded that it was time to go back. Of course I had made my decision to go back before I ever lay down on that table. But, you know, the more I was there, the better I liked it [laughs]. My grandmother didn't take me back through the tunnel or even send me back or ask me to go. She just looked up at me. I expected to go with her. My uncle was the one who brought me back down to the body. But then I got to where the body was, and I looked at the thing, and I for sure didn't want to get in it because it looked pretty much like what it was: void of life. I believe it was covered. It scared me, and I didn't want to look at it. And I knew it would hurt, so

I didn't want to get in. But he kept reasoning with me. He said: 'Like diving into a swimming pool, just jump in.'

'No.'

'What about the children?'

"You know what, the children will be fine [laughs].' And he goes, 'Honey, you got to go.'

'No.' He pushed me; he gave me a little help there. It's taken a long time, but I think I'm ready to forgive him for that [laughs].

I saw the body jump.... And then he pushed me, and I felt it chill me inside. I returned to my body. It was like diving into a pool of ice water.... It hurt!

When I came back, and I was still under general anesthesia in the operating theater, they were playing "Hotel California," and the line was "You can check out anytime you like, but you can never leave." I mentioned [later] to Dr. Brown that that was incredibly insensitive, and he told me that I needed to sleep more [laughter]. When I regained consciousness, I was still on the respirator."

Pam concludes her account by saying, "I think death is an illusion. I think death is a really nasty, bad lie."

Dr. Sadom participated in Pam's surgery and commented on her story: "I found that what she saw from her out-of-body experience seemingly corresponded very accurately to what had actually occurred. She looked at the bone-saw that was being used to cut open her skull. It indeed does resemble an electric toothbrush.... And there was some conversation at the time between the doctors and Pam accurately recalled hearing that conversation...."

Dr. Spetzler also weighed in: "I don't think that the observations she made were based on what she experienced as she went into the operating theatre. They were just not available to her. For example, the drill and so on, those things are all covered up. They aren't visible; they were inside their packages. You really don't begin to open until the patient is completely asleep so that you maintain a sterile environment.... I find it inconceivable that the normal senses, such as hearing, let alone the fact that she had clicking devices in each ear, that there was any way for her to hear those through normal auditory pathways.... I don't have an explanation for it. I don't know how it's possible for it to happen, considering the physiological state she was in. At the same time, I have seen so many things that I can't explain that I don't want to be so arrogant as to be able to say that there's no way it can happen."

The bright white light often reported during NDEs also appears in studies of voluntary OBEs. Here's another testimony William Buhlman received for his international out-of-body survey:

> *I have traveled out-of-body on several occasions, but there was one time when I felt totally different. I was flying through the clouds, and past the clouds into a bright white-yellowish light. As I was going to the light, an extreme peacefulness came over me. I felt very light...I also felt an overwhelming amount of love come through me. It was something I had never felt before. The light was so bright, I felt like squinting when I saw it...but I just kept my eyes open and looked straight at the light and saw that it was not hurting my eyes. Looking around, I heard voices in the distance. I saw my grandparents, who had died some 10, to 15 years before. I was so very happy to see them. I felt our energies become one. I felt like I was home. I wanted to stay...but they told me that I was to go back, for it was not my time to be with them. Reluctantly, I came back.*

When I came back into my body, it felt as if I was "jump-started" like a jolt of electricity. I was gone at that time for 45 minutes. —Bob T. Mobile, Alabama

According to the Grayson Scale, this self-initiated out-of-body experience would be classified as an NDE. Feelings of peace and joy, the perception of a brilliant and non-blinding bright light, an out-of-body experience and meeting deceased relatives are all elements of the Grayson Scale.

When I asked Tom Campbell about near-death experiences and the deceased family members that a lot of people experience meeting at the white light, he said that these family members are simulated to give the deceased person a sense of familiarity. Campbell believes that a larger consciousness system supervises the growth of every soul or *individuated consciousness unit* as he refers to it. He says when a person "exits a fully immersive physical reality simulation" they are often confused about who they are and where they are. So the larger consciousness system simulates realities to provide a sense of familiarity during this transitional phase.

I asked him how he knew this and he said that he used to work at the "way station" between lives. It is rare that you find a nuclear physicist who believes in reincarnation, perhaps even rarer one who experiences to have worked with people who are transitioning between lives.

He told me, "I have a day job, and I have a night job."

CHAPTER 9:

Unexpected Returns

The case of James Leininger and Dr. Ian Stevenson's research of children's past lives

The idea of reincarnation has been part of religions and mystical traditions for eons. In the 1960s, reincarnation started to become a field of scientific research, particularly with regards to children's past life memories. One of the famous cases that popularized this field of research was that of James Leininger.

James was born in 1998. When James was just 2 years old, he started talking about his experiences as a wartime pilot. His parents, both devout Christians, could not make sense of their son's terrifying memory of dying in an airplane crash.

James told his mother that the airplane he flew was a Corsair, that he took off from a boat called the Natoma, and that he was shot down by the Japanese. He even remembered that he had a friend named Jack Larsen and that the tires of the Corsair airplanes often popped during landings.

Eventually his parents considered the possibility that their son was remembering a past life. They asked him what his name was and he said it was also "James." They did some research and were able to find records of a James Huston, who worked with a Jack Larsen, was stationed on an aircraft carrier called the *U.S.S Natoma Bay*, flew a Corsair, and was shot down by the Japanese in 1945.[11]

Psychiatrist Dr. Ian Stevenson, who died in 2007, was a researcher who studied children's past life memories starting in 1960. Stevenson traveled the world investigating past life memories of children and collected over 2,500 cases suggestive of reincarnation. He interviewed over 2,500 families about their children's past lives.

In one case from India, a 10-year-old boy named Nirmal died of smallpox in April 1950 in a village called Kosi Kalan. On the day of his death he told his mother, "You are not my mother. You are a Jatni. I will go to my mother." When saying this, he pointed in the direction of Indian towns Chhatta and Mathura.

In August 1951, the wife of Sri Brijlal gave birth to her son Prakash in Chhatta. One night, when Prakash was 4 years old, he attempted to run away. When his family stopped him, he said that he belonged to Kosi Kalan. For a month Prakash kept trying to run away, begging his family to let him go back to Kosi Kalan.

Prakash could vividly recall the life of Nirmal, the 10-year-old-boy who had died from smallpox. He remembered the names of Nirmal's family members and friends. Prakash's current family grew increasingly impatient and began beating him and spinning him counter-clockwise on a potter's wheel to make him forget about Nirmal's life.

The brother of Nirmal heard that a boy from Chhatta claimed to be the reincarnation of Nirmal. He told his father and one day, on a business trip to Chhatta, Nirmal's father went to look for the boy. He found him and Prakash immediately recognized him as his father.

---

[11] http://abcnews.go.com/Primetime/Technology/story?id=894217&page=1

Prakash begged his current family to let him go with his previous father, but they did not.

On another occasion, Nirmal's older sister went to visit the boy in Chhatta and Prakash wept with joy when he saw her. He begged his current family again to let him go to Kosi Kalan, and eventually his parents agreed to let him visit.

Upon arriving, Prakash recognized other family members, neighbors, and parts of the house Nirmal had lived in. The visit to Kosi Kalan increased Prakash's longing to live with Nirmal's family, and he once again started running away from home. However, animosity built up between the families because Prakash's family suspected that Nirmal's family was planning on permanently adopting him, and the visits stopped.

Dr. Stevenson visited Prakash one more time, in 1971. Prakash was 20 years old by then, and said he did not think about his previous life any more unless he was asked about it.[12] How such verifiable knowledge could travel from one child to another without any known communication remains a mystery.

Some scientists say Dr. Stevenson was gullible and was deceived by the people he interviewed; others ignore his research altogether. While it's quite possible that not every account Stevenson received was accurate, it also seems rather unlikely that Stevenson was deceived over two thousand times.

I contacted Dr. Stevenson's colleague Jim Tucker to find out more about their research project. We scheduled a Skype call and talked for an hour or so. Tucker was wearing a blue shirt and an orange tie. He

---

[12] Dr. Ian Stevenson, *Twenty Cases Suggestive of Reincarnation*, (University of Virginia Press, 1980)

talked very matter-of-factly about the cases he studied. His attitude was very scientific in the sense that he wasn't trying to prove a belief in past lives, but rather trying to explain the observed phenomena of children's past life memories. Tucker told me that he and Stevenson carefully investigated all the cases they publicized. They never offered to pay the people they interviewed, since compensation would incentivize people to make up stories, and they only interviewed first-hand witnesses.

It is true, however, that Stevenson and Tucker collected more cases in countries in which reincarnation is a widespread belief. Tucker explained that people who believe in reincarnation are more willing to share their stories, noting that in the U.S. you couldn't just walk into a grocery store and ask if anyone has heard about a child remembering a past life.

That said, Stevenson and Tucker also investigated quite a few cases in the U.S. and Europe. Tucker told me about the case of a young boy remembering his grandfather's life in New York.

In 1992, John McConnell was working as a security guard in New York. One night after work, he stopped by an electronics store and found two people robbing it. He pulled out his pistol, but before he got a chance to fire a bullet he was shot six times. His left lung, heart, and main pulmonary artery were punctured, and McConnell did not survive.

Five years after John McConnell's death, his daughter Doreen gave birth to her son William. William was born with a medical condition called pulmonary valve atresia. The valve of his pulmonary artery and one of the chambers of his heart had not formed properly. William underwent surgery and had to take medication indefinitely.

One time William was being difficult, and Doreen threatened to spank him, to which he reportedly said that he never hit her when he was her dad.

Another time William asked his mother what their cat's name was when he was her dad. Doreen was very startled, but told William she grew up with a white cat named Boston in the family. When Doreen told William the cat's name, William remembered calling the cat Boss. Boss was indeed Boston's nickname — a nickname that only John McConnell used for the pet.

These strange interactions made Doreen question whether her son could be the reincarnation of her father — especially when William told his mom that he died on Thursday and was born on Tuesday, which was the day her dad died and the day William was born. Eventually Doreen found comfort in the idea that her father had come back into her life.

One day, Doreen asked William if he remembered anything about the five years between John's death and William's birth. William responded: "When you die, you don't go right to heaven. You go to different levels—here, then here, then here," he said as he moved his hand up. William went on to say that animals also go to heaven and can be reborn as human beings. He said he talked to God and asked to be born back into the world.

Doreen told Tucker that William reminds her of her father in many ways. He is a lover of books, a non-stop talker, good with his hands, and especially reminds her of John when he says, "Don't worry, Mom, I'll take care of you."[13]

---

[13] Jim Tucker, *Life Before Life*, (St. Martin's Griffin; 1st edition April 1, 2008)

The sheer amount of cases that have been collected in the field of children's past life memories make one wonder if perhaps Hinduism and Buddhism had captured an existential truth that Western culture has forgotten. Maybe Tom Campbell was right when he said that consciousness is not a product of time and space, but plays a game in which it constructs characters in virtual realities.

CHAPTER 10:

Exhuming the Past

Dr. Brian Weiss, past life regression therapy

Perhaps one of the most controversial fields of research in psychology is past life regression therapy. There is no empirical evidence in this field of research, there are hardly any verifiable facts, and there is a lot of outlandish information. That said, it is an interesting topic to explore in the context of OBEs, NDEs, and children's past life memories, since elements of each of these fields of research also appear in past life regression therapy.

In past life regression therapy, the research subject or client is usually put into a trance through hypnosis. While in this unconscious — or perhaps superconscious — state, the client has a conversation with the psychologist.

In his 1988 book *Many Lives, Many Masters*, psychologist Brian Weiss tells the story of how he accidentally came across past life regression therapy. Dr. Weiss had previously followed a traditional career path, graduating from Yale University and later becoming head of psychiatry at Mount Sinai Medical Center in New York, and had no interest or belief in the paranormal.

However, in 1980, one of his patients started talking about past lives. Catherine was a young woman with severe anxiety, and Dr.

Weiss tried to treat her with hypnosis sessions. During their sessions, Catherine descended into a deep trance and began to speak about vivid memories that did not seem to fit into any psychological framework.

Weiss documented Catherine's words in his book:

> *I see an old man with a beard. He's one of the healers in the village. He tells you what to do. There is some type of… plague … killing the people. They're not embalming because they're afraid of the disease… I also have some disease from the water. It makes your stomach hurt. The disease is of the bowel and stomach. You lose so much water from the body. I'm by the water to bring more back, but that's what is killing us. I bring the water back. I see my mother and brothers. My father has already died. My brothers are very sick… I'm lying on a pallet with some type of covering.*[14]

Then she seems to remember her own death. She said that she left her body and was drawn towards a white light where she met with spirit entities. She went on to remember dozens of different lives that all ended similarly.

All of this was very hard to digest for Dr. Weiss. Initially, he thought that Catherine might just have a very vivid imagination, but he began to question this hypothesis when Catherine revealed information she could not have gained through her five senses.

As Dr. Weiss preferred to maintain formal relationships with his patients, he never spoke to them about his private life. During one of their sessions, Catherine described dying in another past life and floating into the white light. Then she shocked Weiss by continuing:

---

[14] Brian L. Weiss, Many Lives, Many Masters, (Fireside July 15, 1988)

"Your father is here, and your son, who is a small child. Your father says you will know him because his name is Avrom, and your daughter is named after him. Also, his death was due to his heart. Your son's heart was also important, for it was backward, like a chicken's. He made a great sacrifice for you out of his love. His soul is very advanced ... His death satisfied his parents' debts. Also, he wanted to show you that medicine could only go so far, that its scope is very limited."

Dr. Weiss was speechless. How could she know all these things? His father's Hebrew name, his father's heart attack, that Weiss named his daughter after his father, about his son's death, his son's backward heart? His son indeed died as an infant from a rare heart disease in which his pulmonary veins entered his heart on the wrong side. Weiss said that Catherine's voice had changed while she revealed this information. He got the feeling that another entity was speaking through her vocal cords.

"Who is there?" asked Weiss. "Who tells you these things?"

"The Masters," she whispered, "the Master Spirits tell me."

When Dr. Weiss asked her after a hypnosis session what "the Masters" meant to her, she thought he was talking about the golf tournament. She had no recollection of talking to or about "the Master Spirits" when she was awake.

During a different session, Catherine remembered living a simple life as a farmer. After remembering a painful death, she again described the white light:

"I am aware of a bright light. It's wonderful; you get energy from this light." Minutes passed in silence before she added in a loud, raspy voice: "Our task is to learn, to become God-like through knowledge."

The things Catherine said during her therapy sessions changed Dr. Weiss's perspective on life and also helped to heal Catherine's anxiety.

Dr. Weiss noted that Catherine's fears began to diminish as she relived what he eventually interpreted to be traumatic past lives. He proposed the hypothesis that some of our fears originate in past lives or traumatic death experiences. Many of the experiences Catherine relived were traumatic deaths; as she re-lived them, she released her fears.

Dr. Weiss' experience with Catherine caused him to turn his back on traditional psychology, devoting the remainder of his career to expanding the framework of psychology and popularizing past life regression therapy.

Dr. Michael Newton is another hypnotherapist who accidentally stumbled across past life regression therapy. He was also a traditional psychologist until a client came to him with chronic pain. The client said that he had seen numerous doctors and was told over and over again that his symptoms were psychosomatic and that he should see a psychologist. Newton put him into a trance and asked him about the pain, at which point the client claimed to remember being stabbed with a bayonet in a battle. This incident started Dr. Newton's career as a past life regression therapist and author.

Dr. Newton's approach to the subject is unique. It sometimes appears that he leaves behind his role as a therapist and asks questions as a nosy reporter would. This makes his books especially entertaining, but also hard to believe. I did not manage to interview Dr. Newton because his secretary told me that he has retired and no longer does interviews. It is easier to tell someone's motivations when you meet them in person, and because I didn't have to chance to speak to Newton directly I don't have a personal opinion of whether his research is legit or not. Although the case studies he published seem to support the ideas I presented in previous chapters, I do think it is important to look at them with an open but skeptical mind. He could be a groundbreaking researcher or just someone trying to capitalize on hot topics in the New Age community.

In his book *Journeys of Souls*, Dr. Newton compiled 39 case studies. Although none of these provide empirical evidence, it is striking that many of these case studies sound like memories of near-death experiences. Here is an excerpt from what Newton claims to be a past death memory:

> *S: At first ... it was very bright ... close to the Earth ... now it's a little darker because I have gone into a tunnel.*
> *Dr. N: Describe this tunnel for me.*
> *S: It's a ... hollow, dim vent ... and there is a small circle of light at the other end.*
> *Dr. N: Okay, what happens to you next?*
> *S: I feel a tugging ... a gentle pulling ... I think I'm supposed to drift through this tunnel ... and I do. It is more gray than dark now, because the bright circle is expanding in front of me. It's as if ... (client stops)*
> *Dr. N: Go on.*
> *S: I'm being summoned forward ... the circle of light grows very wide and ... I'm out of the tunnel. There is a ... cloudy brightness ... a light fog. I'm filtering through it.*
> *Dr. N: As you leave the tunnel, what else stands out in your mind besides the lack of absolute visual clarity?*
> *S: (subject lowers voice) It's so ... still ... it is such a quiet place to be in."*[15]

[15] Dr Michael Newton, *Journey of Soul*, (Llewellyn Publications; 1st edition July 1994)

In another of Dr. Newton's past life regression case studies, the subject describes entering what appears to be a non-physical reality:

Dr. N: As you move further away from the tunnel, describe what you see around you in as much detail as possible.

S: Things are … layered.

Dr. N: Layered in what way?

S: Umm, sort of like … a cake.

Dr. N: Using a cake as a model, explain what you mean?

S: I mean some cakes have small tops and are wide at the bottom. It's not like that when I get through the tunnel. I see layers … levels of light … they appear to me to be … translucent … indented …

Dr. N: Do you see the spirit world here as made up of a solid structure?

S: That's what I'm trying to explain. It's not solid, although you might think so at first. It's layered—the levels of light are all woven together in … stratified threads. I don't want to make it sound like things are not symmetrical—they are. But I see variations in thickness and color refraction in the layers. They also shift back and forth. I always notice this as I travel away from Earth.

Dr. N: Why do you think this is so?

S: I don't know. I didn't design it.

Dr. N: From your description, I picture the spirit world as a huge tier with layers of shaded sections from top to bottom.

S: Yes, and the sections are rounded—they curve away from me as I float through them.

Dr. N: From your position of observation, can you tell me about the different colors of the layers?

S: I didn't say the layers had any major color tones. They are all variations of white. It is lighter ... brighter where I'm going, than where I have been. Around me now is a hazy whiteness which was much brighter than the tunnel.

Dr. N: As you float through these spiritual layers, is your soul moving up or down?

S: Neither. I am moving across.

Dr. N: Well, then, do you see the spirit world at this moment in linear dimensions of lines and angles as you move across?

S: (pause) For me it is ... mostly sweeping, non-material energy which is broken into layers by light and dark color variations. I think something is ... pulling me into my proper level of travel..."[16]

This description of the spirit world or the nonphysical reality is very similar to what Robert Monroe experienced during his out-of-body travels. Monroe talked about it in his book *Far Journeys*:

We drifted beyond the outer edge of the rings that make up what I was later to realize were the Belief System Territories, parts of the (M) Field spectrum adjacent to the Earth Life System where many Human Minds reside after completing physical life experiences. (M field spectrum is thought energy.) We could perceive the Earth in the center with semitransparent radiant globes around it, each larger and thinner as the distance increased. It took some effort

[16] Dr Michael Newton, (July 1994)

to recognize that we were 'seeing' the nonphysical energies in the structure rather than electrons and molecules.[17]

Later on, Monroe spent many years exploring these layers in nonphysical realities, which he called "the belief system rings." He calls them belief systems because, according to him, within them are all kinds of experiential realities or virtual realities that are inhabited by souls that share a belief. According to Monroe, the belief system rings are organized by a spectrum ranging from fear to love. The darker layers closer to the planet are inhabited by more fearful souls, Monroe claims, while the outer layers are inhabited by older souls that are further along the path. This would imply that the spirit world is organized and every soul has a particular place.

The case studies of Dr. Michael Newton suggest something similar. Most of his clients said that after death they were brought to a place where they felt they belonged. Sometimes guides or relatives meet them and bring them there. Other times they describe being pulled by invisible strings, or being guided by a wave of music and the feeling of love. This type of experience is also common during NDEs.

Dr. Newton claims to have had the following conversation during a hypnosis session:

Dr. N: What do you see around you?

S: It's as if ... I'm drifting along on ... pure white sand ... which is shifting around me ... and I'm under a giant beach umbrella—with brightly colored panels—all vaporized, but banded together, too ...

Dr. N: Is anyone here to meet you?

---

[17] Robert Monroe, *Far Journeys*, (Harmony; Reissue edition September 14, 1987)

S: (pause) I … thought I was alone … but … (a long hesitation) in the distance … uh … light … moving fast towards me … oh, my gosh!

Dr. N: What is it?

S: (excitedly) Uncle Charlie! (loudly) Uncle Charlie, I'm over here!

Dr. N: Why does this particular person come to meet you first?

S: (in a preoccupied far-off voice) Uncle Charlie, I've missed you so much.

Dr. N: (I repeat my question)

S: Because, of all my relatives, I loved him more than anybody. He died when I was a child and I never got over it. (on a Nebraska farm in this subject's most immediate past life)

Dr. N: How do you know it's Uncle Charlie? Does he have features you recognize?

S: (subject is squirming with excitement in her chair) Sure, sure— just as I remember him—jolly, kind, lovable—he is next to me. (chuckles)

S: He is smiling and holding out his hand to me …

Dr. N: Does this mean he has a body of some sort with hands?

S: (laughs) Well, yes and no. I'm floating around and so is he. It's … in my mind … he is showing all of himself to me … and what I am most aware of … is his hand stretched out to me.

Dr. N: Why is he holding out his hand to you in a materialized way?

S: (pause) To … comfort me … to lead me … further into the light.

Dr. N: And what do you do?

S: I'm going with him and we are thinking about the good times we spent together playing in the hay on the farm.

Dr. N: And he is letting you see all this in your mind so you will know who he is?

S: Yes … as I knew him in my last life … so I won't be afraid. He knows I am still a little shocked over my death. (subject had died suddenly in an automobile accident)

Dr. N: Then, right after death, no matter how many deaths we may have experienced in other lives, it is possible to be a little fearful until we get used to the spirit world again?

S: It's not really fear—that's wrong—more like I'm apprehensive, maybe. It varies for me each time. The car crash caught me unprepared. I'm still a little mixed up.

Dr. N: All right, let's go forward a bit more. What is Uncle Charlie doing now?

S: He is taking me to the … place I should go …

Dr. N: On the count of three, let's go there. One—two—three! Tell me what is happening.

S: (long pause) There … are … other people around … and they look … friendly … as I approach … they seem to want me to join them …

Dr. N: Continue to move towards them. Do you get the impression they might be waiting for you?

S: (recognition) Yes! In fact, I realize I have been with them before … (pause) No, don't go!

Dr. N: What's happening now?

S: (very upset) Uncle Charlie is leaving me. Why is he going away?

Dr. N: Look deeply with your inner mind. You must realize why Uncle Charlie is leaving you at this point?

S: (more relaxed, but with regret) Yes ... he stays in a ... different place than I do ... he just came to meet me ... to bring me here."[18]

Newton interprets the place the subject is brought to as her soul group. According to Newton, a soul group is a cluster of souls that are working together throughout several lifetimes and help each other reflect on the experiences of their physical lives and also incarnate together as friends or family. But it should also be noted that in this case study Dr. Newton doesn't behave very objectively. He appears to be guiding his patient very actively, which in my opinion makes the study less believable.

On the other hand, I have come across dozens of transcripts of past life regression therapy sessions that all talk about the educational settings of their soul groups. Maybe the rings Monroe talks about and the layers Dr. Newton's patients describe are like multiple school houses piled on top of each other, bringing souls together who are working through similar growth experiences. Perhaps the different gradients of white represent the collective consciousness quality within these fields or consensus realities. The sheer amount of similar subjective experiences makes one wonder if there could be some truth to these ideas.

Experienced out-of-body explorers like Buhlman and Monroe describe nonphysical realities as vast collections of thought-created

---

[18] Dr Michael Newton, (July 1994)

consensus-realities. Buhlman and Monroe use the term "consensus reality" to describe an experiential reality created through the belief of multiple conscious beings.

In his book called *Far Journeys*, Monroe writes about visiting a religious consensus-reality during an out-of-body experience. During this experience, he claimed to be traveling with a non-material being he called BB. Monroe recounted the experience:

> *I decided we were deep enough into the outer ring, so without an ident (a specific location), I pulled to an easy stop. The haze was less dense, and the shape of buildings, irregularly spaced, began to take form; each a suitable distance from another, each different in design either extremely or slightly. Many seemed to be constructed of stone, most were equipped with spires, steeples, domes, or towers in various configurations, some had elaborate stained-glass circular windows. We moved down near the front of the nearest building. As we did, a woman emerged from the front door and descended the wide steps. When she reached the last step, she looked up and stopped short. There was no fear in her eyes, just uncertainty. I decided to set her straight.*
>
> *'We don't bite.' She responded immediately,*
>
> *'I didn't expect you to. I was trying to decide where to refer you. We have so many committees. You don't act like a newcomer.' I smiled at her,*
>
> *'We're just visiting.'*
>
> *'Our minister says there is no such person as a visitor here.' She replied confidently,*

'You wouldn't have found us if you didn't have our faith. It's really all right to be a newcomer. I'll take you to Thelma. She's in charge of the welcoming committee.' I smiled.

'No, thanks just the same, we're only passing through.' She looked puzzled.

'...Isn't it wonderful to know, really know you have everlasting life?' She stretched her arms outward from her side.

'Oh, I remember so well how I felt when I died and they brought me here. How I harbored little secret doubts, and I do understand how you feel. Sunday school and the indoctrination classes will clear those up for you. Don't worry. It's just unusual that you arrived here on your own.'

'Then this isn't heaven? This isn't where God is?' She laughed lightly.

'That's exactly what I asked when I was a newcomer. Don't be disappointed. We're only at the gates of heaven. Our minister, Dr. Fortune, preaches sermons about it every Sunday. I must confess they're quite different from the ones Reverend Wilson used to preach back when I was living physically, in Lexington.'

'Are you going back?' She frowned,

'You mean back into a physical life?'

'I guess that's what I mean.' She was thoughtful for a moment,

'I don't know. Dr. Fortune has sermons about it. He says when you leave here, you can go back again or you can go somewhere else.'[19]

Skeptics would not consider out-of-body experiences or past life regression therapy a source of objective evidence. At the same time, the nature of the subject might be inherently subjective since by definition we cannot put physical measurement devices into a nonphysical reality. It is deeply intriguing that so many subjective experiences speak of a similar dimension of reality. Why is it that NDEs, OBEs, and past life regression therapy all result in very similar case studies? If reincarnation is part of our soul's growth process, it would mean that evolution extends beyond the physical reality, an idea we'll explore next.

[19] Robert Monroe, *Far Journeys*, (Harmony; Reissue edition September 14, 1987)

CHAPTER 11:
Rational vs Experiential Knowledge

A modern interpretation of reincarnation

For reincarnation to make sense, a soul needs to be an information network that converts specific experiences into a different form of memory, forgetting the details but retaining the essence.

Although consciousness is not physical, we can imagine it as a malleable substance that gradually takes on a form defined by tendencies or habits. So the shapes into which consciousness grows are habits. The instinct-driven behavior patterns of physical beings are like containers into which consciousness is poured. Repeatedly experiencing the dynamic interaction between environment and instincts, causes tendencies to develop. These tendencies are the non-specific memories of the experiences a soul has had.

Such tendencies aren't necessarily created purely by physical experiences. A soul could also be starting its incarnation circles with preexisting tendencies. For example, in the Digital Evolution Lab's Avida project we discussed back in Chapter 1, programmers first create organisms and then let them evolve by relating to each other. It is possible that our souls are also created by an intelligence beyond our physical reality. But regardless of the soul's origin, if it is possible for it to develop and grow through the experience of inhabiting different bodies, then there must be something within the soul that retains the essence of a life but forgets the details. I think this essence is the habitual interaction with the psyche of a physical creature.

If you do something over and over again, you can acquire a skill without understanding the process rationally or vice versa. You could be an art historian without knowing how to paint, for example. Rational knowledge consists of the things we have learned intellectually, while experiential knowledge isn't based on remembered facts. I believe that this experiential knowledge is what our soul retains, even when a specific life is forgotten.

When I was trying to learn how to surf, it became clear to me that it didn't matter if I intellectually understood the process. I realized that what I was learning could only be learned by trying to jump on waves over and over again. I did not have to think about it, but I learned when to paddle, when to stand up and how to distribute my weight to keep the board horizontal; it became a habit. I am still a very bad surfer, but while attempting to acquire this skill I clearly noticed the difference between experiential knowledge and rational knowledge.

Operating a body and balancing the emotions of a physical being is a skill. Souls need to interact with the human psyche over and over again to learn how to balance all the psychological forces that affect our human experience. Riding on the psychological waves of the human experience is like surfing — you have to do it over and over again.

From this perspective, the concept of reincarnation begins to make more sense. What we are gaining from the human experience is not defined by specific memories, but by the habits we develop. This hypothesis would also explain why some children are born with unbelievable talent far beyond their years — such talents could be seen as experiential knowledge. But experiential knowledge doesn't just apply to learned skills. It is not just about habits of action, it also relates to habits of being.

The way we think and feel is a habit, and this habit is also experiential knowledge. It defines you far more than the rational knowledge you have about yourself. You could forget what your name is, where you live, and who you are married to, and still maintain your identity. Your true identity is defined by how you interact with the psychological infrastructure of a human being, not by the human identity you create in the process. If you spend many lifetimes being a teacher, you might learn how to use the intellect in a very precise way and how to explain complex ideas to other people. Your soul's habit will be a teacher personality, but it is not defined by any one of the teacher personalities you created. The more experience we gather, the more complex the habit-based data bank of a soul becomes.

CHAPTER 12:

The Soul As a Relational Network

A modern interpretation of the soul

Today we know that a human being is made up of trillions of cells. If our soul can exist independent of our bodies, how does it interact with our cells?

Conceiving of a person's will as a product of the will of trillions of cells might appear to be in contradiction with the concept of a soul with free will. How could we have souls with free will if our will is supposed to be a product of our cells?

The soul is not the product of the will of our cells; the soul is the relationship between the will of our cells. If the soul was the product of our cells' will, we would be directed by the needs of our cells. We do feel the need of our cells when we are hungry or horny, but instincts don't define us. There are people who fast and people who are celibate. We are not victims of human nature. We experience the will of our cells, but we can choose how to deal with our instincts, how to put our cells into relationships with each other. We are a choosing whole with the ability to reconfigure its parts.

If you choose to meditate or to work out in the park, you create different relationships between your cells. If you choose to meditate, your heart will beat much slower; when you work out your heart will beat faster. These are just the obvious differences, but if you closely study the activity of every cell, you find that these two choices have a completely different effect on your biochemistry and neurological activity. The cells themselves are not choosing to work out; it is the relational infrastructure between them that is making the choices.

A relational infrastructure is a network of relationships. Your soul manifests itself in a body in the form of the relationship between all of your cells. You are the consciousness which is affecting your cells' choices by modifying their probability distributions with your intention. So if you intend to do something, it raises the probability of your cells to make choices that correspond to your intent. When you want to open your mouth, the nonphysical you affects the probability distributions of the conscious agents represented by the cells in your jaw muscles.

Rupert Sheldrake also proposes this hypothesis in relation to nestled morphogenetic fields. As we discussed in Chapter 5, he believes that every physical structure represents a morphogenetic field that contains information about this structure. In this understanding, the cells in your body are linked to an information field which contains the information of all cells of that type that exist or have ever existed.

However, the organs constructed through the collaboration of cells are also linked to an information field which subsumes that of the individual cells. He believes that these fields interact by affecting each other's probability distributions. So the morphogenetic field of the organs can guide the activity of the cells by diminishing or amplifying certain probable choices. He believes that this nested pattern of information fields extends in both directions, and even subsumes the field that I call the human soul.

When we say that souls are affecting bodies by modifying the probable choices of the conscious agents which are represented by cells and organs, we are saying the same thing as Rupert Sheldrake. "Conscious agents" is Donald D. Hoffman's preferred term, but a conscious agent is also an immaterial energy which arranges the information of "physical" structures. The term *soul* also serves the same purpose. A soul is a large network of nestled morphogenetic fields or conscious agents. These agents are all one, but differ in their habituated behaviors. They are all driven by the same conscious energy that has manifested itself in different repetitive motions, on different levels of magnification.

The idea that the soul is a relational network of conscious agents or information fields can coexist with the idea that our bodies are communities of cells. Hoffman would say the community of cells is the interface of conscious agents. Cells are the virtual representation of an immaterial energy. Ultimately there is no material or immaterial, since matter is the interface of consciousness and there is only a difference in data structures. So cells are the data structure within the framework of time and space, but they are being operated by consciousness.

When you choose to incarnate into a body, you are choosing to use the time and space interface to put conscious agents into a relationship with each other. You engage the conscious agents that are represented by your brain cells in a way that is characteristic to your individuality or your habits of thinking and feeling.

As you incarnate over and over again, your soul learns to create more harmony in the context of a complex structure of conscious agents, or cells and their subsystems. Creating harmony in networks of conscious agents is an integrative process explored (with different language and terms) by the teachings of many religions and mystical traditions. So this extended version of the simulation hypothesis views spiritual teachings as programs we can use to create coherence in the complex systems of nestled conscious agents. This might seem unnecessarily convoluted, but the realization that spiritual teachings can be understood logically is a valuable bridge between mysticism and science. This hypothesis is backed up by meditation research that investigates the physical and emotional health benefits of meditation, since health is coherence in regards to the relationships between cells.

Emotional issues that you had in a past life might re-emerge if you have not found integrative solutions, because your soul has not managed to grow beyond a certain point of complexity without losing its willpower to a survival-based drive. In other words, your emotional health makes you aware of your soul's progress of manifesting harmony within the rule set of this physical reality simulation and the survival-based drives you inherit.

We can choose to follow the fear-based drives that have evolved within this reality, or we can choose to embody the harmony of our soul. In the context of modern human civilization, survival-based drives produce stress hormones that are no longer translated into physical activity (for example, it's a much different thing to face a predator in the wilderness than it is to worry about an upcoming rent payment or job interview).

A lot of illness is due to chronic stress. However, the soul has the power to override survival-based drives if it habitually confronts them and creates new neurological pathways through which our cells can communicate in healthier ways. This is what spiritual or personal growth is: a rewiring of the neurological pathways and the soul's growth beyond its habituated way of relating to the human psyche.

We will explore this concept in greater detail in chapter 19, but for now I just want to point out that there is a way to conceive of the soul in terms that complement the findings of science and create a definition for spiritual growth. Spiritual growth or spiritual evolution is the soul learning to create more harmony within complex systems. In a broader sense, consciousness learns to organize the information structure we call matter.

CHAPTER 13:

Three Types of Evolution

Physical evolution, cultural evolution, and spiritual evolution.

The logical consequence of reincarnation is that there are three different types of evolution that all intersect in our perceived reality: physical evolution, cultural evolution, and spiritual evolution. I don't want to traffic in vague New Age terminology, but I do think it is important to build the bridge between thought systems that have previously attempted to conceptualize the growth of the soul. At the same time, I hope that traditional scientists can use the models of Tom Campbell, Donald D. Hoffman, and Rupert Sheldrake to get a larger perspective of what evolution could be and can look past the often-dismissive connotation of the term "spiritual."

Physical evolution, or Darwinian evolution, is the process that changes the DNA of organisms through mutation and natural selection.

Cultural evolution has its roots in Darwinian evolution, but with the growth of civilization, cultural evolution has become driven by complex socioeconomic forces not based on immediate survival concerns.

Spiritual evolution is the learning process of the soul or consciousness and is facilitated by the interaction with physical and cultural evolution.

Within every human being, these three types of evolution interact and create challenges from which the soul can learn, or expand its ability to create coherence within complex free will systems. Souls incarnate to integrate the obstacles of the human being's psychological infrastructure, the obstacles of cultural conditioning, and the soul's accumulated behavior patterns. At its core, the soul's growth is driven by the conflict between a love-based spirit and an experiential reality that is evolving based on survival of the fittest.

Cultural evolution, including the evolution of economic and political systems, is also driven by conflict. The soul's desire is to manifest love, but the human emotions we experience are designed to ensure the survival of our species. In this understanding, we can consider the ego to be a management system or a behavioral network that deals with these survival drives. The ego and the soul are often in conflict. That's why not all things in this perceived reality are aligned with the consciousness modality of love.

Some people are motivated by greed, which is created by the ego's attempt to deal with the fear of not getting enough. Greed or other fear-based motivations are related to the concept of survival of the fittest. The fear of starving prevented early humans from starving, but now these drives have outlived their usefulness, causing imbalances within society and ourselves.

As we deal with the behavior patterns that have evolved here on Earth, we struggle because our awareness wants more harmony than this planet provides. In this struggle, we create social norms, religions, economic systems, institutions and school systems, all of which get passed down to the next generation. Every individual who is born into this reality needs to deal with how our ancestors made sense of the world, with the institutions they created, the social norms they reinforced, and even the epigenetic changes they made to their DNA.

The more we reincarnate into this reality and deal with these forces, the better we get at manifesting our souls' intention despite the resistance of culture and psyche. It is hard to make sense of the world with a human brain, particularly in a culture that has not yet evolved to the point where spiritual growth is taught in school. But the difficulties of physical life might actually be the reason why souls incarnate. If evolution extends beyond the physical reality, then we could view our physical and cultural predispositions as resistance that fuels the evolution of our soul. From this perspective, our difficulties on this planet are just as necessary as weights at the gym.

CHAPTER 14:

Self-actualization

Religious, spiritual, psychological, and practical interpretations of self actualization

My friend Mike once told me about a spontaneous mystical experience that he had when he gave a homeless woman some money. We were playing a show in Bergen, Norway. A few hours before doors opened, he walked through the streets to look for some food. An elderly woman was sitting on the sidewalk asking for money. He gave her a relatively large bill and she stood up in complete disbelief and hugged him with gratitude.

Mike told me he felt a divine presence in that moment. The feeling was so strong that it was as if the reality around him had dissolved and he was taken into another realm. As he experienced this divine presence, it became clear to him that his purpose and responsibility was to self-actualize, to fulfill his potential, to become this immense love.

Self-actualization refers to a process. I like to think about this process in terms of crystallization, which occurs when less organized patterns become more organized.

For example, a quartz crystal is an iterative tetrahedral pattern of silicon and oxygen atoms. There are four oxygen atoms and one silicon atom in each iteration. The four oxygen atoms are positioned on the corners of a three-dimensional triangle known as a tetrahedron, and the silicon is in the middle. The corners of each tetrahedron touches another one, and together they form three-dimensional hexagons. This pattern repeats over and over again. A quartz crystal is an example of an organized pattern that two different elements can create. Over the course of millions of years, disorganized atoms come together and form highly organized patterns.

There are also more complex crystals with more elements. In an abstract sense, anything can be a crystal as long as it consists of synchronized patterns. In Chapter 3, we explored how the state of love or bliss results in synchronized brain waves. We can think of this as psychological crystallization, because the growth towards heightened psychological emotions such as bliss is the growth towards coherence and order.

Dr. Richard Davidson demonstrated in an experiment with long-term meditators that heightened emotional states such as compassion and bliss correlate with coherent gamma waves and synchronized brain and heart communication.[20] I believe that this is probably true for anyone who has somehow found inner peace. Just like the different elements in the quartz crystal align to form an ordered pattern, the state of peace or bliss aligns processes within our body.

It is well known that crystals often emerge from a germ. This means that a small cluster of crystallized molecules begins to organize the elements around it, and in the process the crystal grows. We could look at our souls as crystal germs, in the sense that we have an essence that seeks self-actualization.

---

[20] Daniel Goleman & Richard J. Davidson, Altered Traits, (Avery 2017)

The psychologist Abraham Maslow was one of the first Western psychologists who proposed this idea, although he didn't use the same metaphors. Maslow believed that every person has a basic nature or essence that wishes to be actualized. Maslow considered actualized people to be those who had found inner peace and maintained a positive and loving relationship with themselves and others. He believed that the process of becoming actualized is different for every person and the only guideline is this essence and its desire to be expressed.

This idea resembles Hindu and Buddhist philosophy and is also reflected by esoteric Christianity, which refers to the concept as "the Christ within."

Although modern, mainstream Christianity is dramatically different than its early forms, there is a lot of evidence that Christianity used to have an esoteric component used by initiates to expand their consciousness. In his book *Inner Christianity,* the scholar Richard Smoley commented on this: "Both the mystical and esoteric paths are generously represented in the Christian tradition." For example, the 14th-century text known as The Cloud of Unknowing is a teaching that leads the reader to know God through the stillness of the heart. Knowing God through the stillness of the heart is a meditative state of consciousness and a poetic description of a coherent state of consciousness. Perhaps "coherent state of consciousness" is a scientific description for knowing God.

More examples of meditative practices in Christianity appear in Quietism of the 1600s and in Quaker spirituality. I believe that there are many paths that lead to inner peace, and when a practitioner experiences "luminous stillness," it results in the alignment of brain waves, which is a modern way of saying what ancient mystics have long pointed towards.

According to Smoley's book, every major religion has inner teachings and those inner teachings are all very similar. They have their unique terminology and metaphors, but what they have in common is a framework that allows practitioners to expand an aspect of the self, which grows into the experience of a divine connection.

In Buddhism, the growth of the self is often compared to a lotus flower. The lotus flower feeds off the sludge on the bottom of a lake to grow into its potential. In the same way, the self, when it is properly nurtured, transforms the raw materials of the human psyche and grows into its potential. This potential is seen as the Buddha. Buddhists believe that we all have the Buddha's seeds within us and that there are teachings and practices that make these seeds sprout, grow and flower as enlightened beings.

When you let your essence grow into its full potential, it reorganizes your psyche like a lotus flower reorganizes the molecules in the soil, or like a crystal germ reorganizes the minerals of its surroundings.

When something grows towards a higher level of organization, it converts disorganized patterns into more organized patterns. Tom Campbell believes that this is the purpose of the reincarnation system. Souls come into physical existence to actualize their potential by reworking the disharmonious mental and physical structures of a young civilization. Like a plant that grows in soil, a soul grows in the psyche of a human being. The human psyche has evolved here on Earth based on survival of the fittest, which produces instincts and drives that are in opposition with the soul's intention. However, the psyche is also malleable, and the soul has the ability to reorganize these patterns like a crystal germ that reorganizes the raw materials it comes in contact with. We use the information patterns evolution has created and convert them into more harmonious patterns, like a plant that creates a flower out of dirt.

If you spend several years applying the tools of self-actualization, you'll experience less suffering and more harmonious emotions. You will notice a difference in your perception of the world, in the way you affect others, and in the way you feel about yourself. This change is reflected in brain and heart coherence.[21]

However, it is difficult to express one's potential in physical form because the behavior pattern or instincts that have evolved here have the power to override the will of our souls. Emotions like jealousy, competition, and anger can drive us away from becoming self-actualized. When we incarnate, we interact with these difficult emotions and they have the power to shape our beliefs and goals, but we also have the power to rework these emotional patterns and find harmony within all these seemingly conflicting drives.

Meditation teaches us not to react to thoughts and emotions. With practice, this allows psychic energy to flow freely and to form more coherent patterns, which evolve into coherent emotions such as love, gratitude, and compassion. These emotions produce very coherent brain waves, but they are products of more painful emotions. This is why Abraham Maslow said that growth often occurs through tragic experiences and painful emotions. This idea also appears in the work of popular German spiritual teacher Eckhart Tolle, who has said that focusing on your emotional pain without trying to push it away gradually transforms it into peace.

Pain is the raw material out of which the enlightened being grows. In the same way, a seed uses the decomposed materials in the soil to

---

[21] Daniel Goleman & Richard J. Davidson, Altered Traits, (Avery 2017)

grow into a tree, and the crystal germ uses the disorganized minerals around it to create a crystal structure. We use painful emotions to grow into our full potential. All of us are engaged in this process even if we are not aware of it. We are all love-seeking beings who have to struggle with emotions that have evolved on this planet. This struggle is the process of self-actualization.

In many spiritual traditions, self-actualization is referred to as an awakening. Normal consciousness is seen as a type of sleep within which one is driven by sensual desires and psychological needs.

From the perspective of the simulation hypothesis, awakening could be considered the process of recognizing the true nature of oneself within many layers of illusion. But the realization of awakening isn't just rational, it is experiential.

If our bodies are actually our avatars in a fully immersive virtual reality, it follows that who we thought we were is an illusion and the only real thing is our awareness, the part of us that experiences. Perhaps we cannot describe what our awareness is without creating another illusionary belief system or theory, and perhaps this is the paradoxical undertaking of this book.

Spiritual awakening is the process of recognizing our awareness as the only truth we can know with certainty. The nature of our experience is up for interpretation, but the fact that we are aware is not. Even if you are a materialistic scientist and you doubt whether awareness is an illusion or not, that doubt arises within your awareness. The doubting scientist is aware of his doubt, but perhaps he doesn't pay much attention to the fact that he is aware since his awareness is occupied with an analytical thought process.

The way we focus our awareness is a choice. One person might prefer to construct logical theories, while another might prefer to focus on intuitive, or emotional understanding. Everyone approaches the existential questions of life differently, but regardless of how you explore the nature of reality, your awareness is always at the core of your perception. This is something you can test in your own psyche, by asking yourself: "When I believe something is true, am I aware of my belief?"

Every rational exploration, true or false, is driven by awareness. There is always a driving force which determines how to deal with rational information or any type of information. You are also aware of what you taste, smell, see, hear, and feel. Rational computation could be considered to be another sense since it provides your awareness with information. Rational understanding is another layer of information processing which your awareness utilizes.

Awareness is the energy that drives every aspect of your being. It is the engine which drives your sexual desires, your fears, your needs, your beliefs, and everything else that you have experienced. It is always there, but during normal consciousness, we are focused on the external reality, on the things we experience and not on the aspect that experiences.

The exploration of our awareness isn't just an analysis, since an analysis would only serve us at the level of rational thinking. To really get to know one's own awareness, one has to meet awareness without the things it creates, without thoughts, feelings, and beliefs. If we study awakening only on a rational level, it doesn't provide us with experiential knowledge. It is an entirely different thing to think about awakening or to go through the process of learning how to experience our awareness. The difference is that of a scholar and a practitioner. Scholars study a subject not in relation to themselves, while practitioners apply the subject to themselves. Both approaches are useful in the quest of our essence.

We can use ideas and theories, apply them to ourselves, and then see if they need to be modified to better explain our experiences. Every theory about self-actualization is like a key that is only helpful if you take it, open the door and walk in. The room of inner knowledge has many doors; there is even space in the wall to create your own.

Whether you think of self-actualization in terms of psychological crystallization or knowing God through the stillness of your heart doesn't matter. What matters is that you apply these frameworks and become that undefinable thing that sits at the center of your being.

CHAPTER 15:

Bridges

Using modern frameworks to understand ancient spirituality.

The idea that we are here to grow our awareness and awaken to our true natures isn't a new one. Although many of the related terms we use today were popularized by the hippie and New Age movements of the 1960s, the underlying concepts are ancient. The essence of these terms points towards concepts that have been understood by humans for as long as we have been on this planet. There have always been small groups of people who dedicated their energy towards enlightenment or self-actualization, creating unique belief systems to accompany that process.

In 2018, I went to the annual Aniwa event in Tolland, Massachusetts. Aniwa is a gathering of elders from dozens of different indigenous tribes. I attended different ceremonies and listened to their different worldviews. Although the belief systems they shared had evolved in different continents, they all seemed to be circling around the same sequence of perceived truths. This sequence was the outline of a path beyond selfishness and beyond fear, facing the resistance of human nature, and becoming our authentic selves.

The details of these belief systems varied wildly, from ideas about the cosmos to claims about the history of humanity. Although the specifics of these ideas weren't scientific, they seemed to provide an interface to grow towards and express heightened emotions such as gratitude, bliss, love, and compassion. I don't know if these shamans and sages would have agreed with me, but I interpreted their teachings as a system which has evolved to cultivate heightened emotions within a tribe or society that hadn't chosen the path of reason and scientific exploration.

I believe that beings in the nonphysical reality reveal themselves to us in the form that is most understandable to us. If one expects to meet a specific God or guiding spirit, your nonphysical helpers can show themselves in that form if it serves the purpose of your soul's evolution. If MUI theory is correct, form is an interface, and like any interface, form conveys meaning. The objective reality that exists beyond form is constructed of conscious agents — will-based energies that build data-sharing systems. Anything falls in that category, but because these data-sharing systems are all very different, they use different symbols. Just like there are different icons for different programs on your computer screen, there are different forms and different beings within your subjective experience. Each one of these forms convey meaning to you, which allows you to form a relationship with that system, and so the network grows.

This way of thinking might sound completely out of place when discussing ancient spiritual practices. However, it is the only logical way that I have been able to connect to the mysteries of ancient traditions. One might say that this way of thinking is missing the essence of these traditions, but on the other hand, *every* way of thinking is arguably missing the essence of spirituality. This doesn't mean that there isn't a way to logically understand spirituality, it just means that logical understanding cannot capture the profound emotions triggered through spiritual practice, worship, or prayer. Not every person can readily surrender to an experience or practice that doesn't make sense to them, so a logical framework is for many people a prerequisite of a spiritual experience. Once you can logically see the possibility of a transcendental reality, you can allow yourself to have the experience.

This is admittedly a controversial opinion in the spiritual community. In fact, at this very retreat, I had an argument about this issue.

I was sitting at a table with a couple of people and found myself in a conversation with an ex-neuroscientist whom I'll call Alex. He had blue eyes and a reddish-blond beard and spoke with a faint German accent. After occupying a highly respected research position, Alex had given up on the possibility that rational exploration could benefit spirituality.

"It takes too much effort to make other scientists take research seriously that tries to explore the physical effects of spiritual experience," he told me. "People simply follow what they feel to be true. Real change happens on the level of feelings."

I agreed, but I didn't believe that fact should stop the scientific and logical exploration of spiritual growth and spiritual experiences. We chatted for some time, losing ourselves in the labyrinth of words and concepts. However, I still believe that logic, even though it functions on a different dimension, can bring someone to the door of a non-conceptual reality. If the possibility of that reality is logically understood, one can then let go of the fear and step into the place of no mind. In this place, you might experience things that sound like fairy tales, but because you have a rational model for nonphysical and interconnected existence, you can integrate these experiences better and with less fear. But this doesn't mean that you have to believe all the simplified myths that are passed on in the New Age community.

I agree with Alex that real change happens on the level of feelings, but to allow this real change many people need a rational framework that connects the otherworldly with the measurable facts of daily life. I have used the idea of an information-based reality and numinous data structures to build a bridge between mystical experiences and my intellect. This intellectual framework evolved out of personal necessity. The simulation hypothesis was simply the framework that appeared to be most effective at uniting the findings of science with mystical experiences. Now I use it not only to interpret what had happened to me but also to put myself into a position in which I can readily receive the blessings of all kinds of mystical traditions. No matter what the details are or where the belief system has evolved, it resulted because someone tried to make sense of a transcended world through the limited mind of a human being and the limited context of a particular culture.

This is what all spiritual belief systems have in common: they are limited models that point towards a reality that transcends the intellect and separate identity.

With this attitude, I went to a Tibetan Buddhist temple in California. I signed up for a 10-day Drupjen retreat. In Vajrayana Buddhism, Drupjen is the honoring of the lineage and all deities that practice or protect the teaching of the Buddha. I was told that Vajrayana Buddhism is a mixture of shamanism and Buddhism. The essence of Buddhism is the rejection of all worldly things and all manifestations of duality. As I understand it, Vajrayana Buddhism has the same nonduality at its center but doesn't entirely reject the concept of duality.

At the retreat, I talked to several people who told me that Vajrayana Buddhism is a mix between shamanism and Buddhism. A shaman's role is to be a bridge between worlds, to work with the underworld as well as with the heavens. I was told that Vajrayana Buddhists carry on this tradition and believe themselves to be working with both enlightened beings and wrathful deities. In this tradition, wrathful deities are thought of as nonphysical beings that operate under the consciousness modality of anger. Like an organism with different types of cells, Vajrayana Buddhism employs all kinds of beings that supposedly work together to spread and protect the teachings of the Buddha. In the past the monks were not exclusively pacifists; they had to protect the monasteries from robbers, so they saw defense as a necessary part of life. Vajrayana Buddhism is the attempt to integrate the path of enlightenment with the need to protect the teachings even if that meant hand-to-hand combat. The remains of this paradox are the foundation of the Drupjen retreat.

The monks who lived at the retreat center were part of a lineage they claimed was directly connected to the Buddha himself. I was sharing a room with two other men. When I first walked into our room, both of them were sitting on their bunks, mumbling mantras. Later I got to talk to one of them, whom I will call Chen.

Chen was a chubby Chinese Buddhist with short black hair and a long face. He had a very friendly but occasionally pushy way of talking.

"You really should take refuge in the Buddha before the retreat begins," he said.

When I responded that I was confused about what exactly that entailed, he said that it was an initiation ritual. I told him that I would do it if the opportunity presented itself.

He brought me down the hill, to the shrine. In the center was a 9-foot-tall statue of the Buddha surrounded by thousands of different spiritual ornaments and items. A dozen people were running around, preparing the shrine for the ceremony.

Chen was standing next to me and respectfully gestured towards the monks, who were in the back of the room.

"Do you feel called to take refuge with one in particular?"

I didn't. On the contrary, I felt awkwardly put on the spot. Group dynamics and spiritual communities often make me feel uncomfortable. Although Chen tried to keep pushing me to take refuge, I decided to go to the cafeteria and eat instead. Chen followed me and told me that there was one more teacher who wasn't at the shrine, Lama Sonam.

As I was trying to eat my meal in the cafeteria, Chen leaned towards me.

"Lama Sonam just walked in," he whispered reverentially. "Maybe you want to take refuge with him?"

I turned around and looked into the eyes of an older Tibetan man. He emanated such a powerful feeling of love that the discomfort I had previously experienced left my chest. Without thinking about it, I quickly told Chen I would take refuge with him.

Chen smiled, got up and introduced us. I found talking to Lama Sonam to be very easy. He asked me a few questions and seemed to have a lighthearted laugh in response to nearly everything. Chen asked him about the possibility of me taking refuge with him and arranged for me to meet him after dinner.

Later in the evening, I found myself sitting on a pillow across from Lama Sonam, who had a bunch of old paper scrolls arranged in front of him. Without explaining anything to me, he started reading from the scrolls. They were in Tibetan, and occasionally he would pause and ask me to repeat a few words. I felt an intense shivering energy envelope my body and my perception of reality started to change, as if I had taken a drug. At some point, this energy became overwhelming, almost unbearable. But I stayed calm and listened to him attentively. It seemed like I was signing some type of contract, and although I didn't understand the words, it felt like I knew exactly what was going on. It felt like I was dedicating myself to the path of enlightenment and allowing all the spirits that were part of this lineage to guide me and push me forward towards my highest potential. I don't know if this was true, but this was how my emotions made sense of the situation.

I was ready to have this emotional experience because I saw the lineage that Lama Sonam was part of as a system that had been growing for thousands of years. I saw him as a physical representation of something that was very vast, but invisible. His body seemed to me like the tip of an iceberg with most of its mass out of sight.

I believe that nonphysical realities are somewhat analogous to an internet in which IP addresses are passed on through physical encounters. You cannot visit a website if you don't have a web address. In the same way, I believe that when you meet someone in our physical reality who is connected to a many nonphysical intelligences, you are introduced to the network that this person is a part of.

This thought gave me shivers through my entire body and in my mind I said, "Thank you! Thank you for having done what you have done." I felt what it must have been like to leave your family and your home to pursue enlightenment in a lonely cave in the mountains. Or to join a monastery and face all the pain that comes with the abandonment of the material world. I felt the motivation behind these actions. I sensed a deep belief in the good of all and an iron determination stronger than any fear. These thoughts and feelings rushed through my body while Lama Sonam recited his prayers with a loving gaze.

It was hard not to break down crying. The emotions and realizations that inhabited my body were filled with incredible beauty, sadness, acceptance, anger, compassion, love, and patience. The emotions I was feeling weren't one emotion, but a buzzing river of all the gifts of the human experience. These emotions weren't separate, either; they were all distilled into one golden teardrop of joy.

After the ceremony was done and Lama Sonam gave me a Tibetan name, he also told me to steer clear of secularism. He said that there are many paths and that my path is to follow many paths.

CHAPTER 16:

Field Tests (II)

Visiting my great grandfather's tribe

The next summer, I was on tour with my band. I had a few days off and found myself close to a Native American reservation between Canada and Michigan. The reservation was on Walpole Island, and when I looked it up I found out that several tribes lived there, including the Potawatomi. My great-grandfather was from the Potawatomi, but because of cultural discrimination he had distanced himself from his lineage, and my family had lost touch with our ancestors.

I'd always felt intrigued by Native American culture and by my great-grandfather's legacy. So I drove to Walpole Island in hopes of meeting someone from my ancestors' tribe.

Walpole Island is located in the mouth of the St. Clair River on Lake St. Clair. The island consists mostly of forests and swamps. Just under 2,000 Native people live on the island. Their houses are small and modest, and many of them are old and falling apart.

As I drove along the island's dirt roads, I saw a man cutting branches off a tree on his front lawn. Later I found out that his name was Press. He had dark red skin and black hair that was short on the sides and long in the back. I stopped the car and asked him where

I could make a fire to honor my great-grandfather. I explained that my great grandfather was from the Potawatomi tribe and I wanted to reconnect with my ancestors.

Press was surprised, but also delighted. As it happened, he was also from the Potawatomi tribe. After we talked for a bit, he suggested a fireplace on the other side of the island might be a good place to honor my great-grandfather. I told him I'd check it out, and asked if I could cook him lunch the next day, an offer he accepted with pleasure.

I left his house and headed to the other side of the island. There I stopped to ask for directions and had a similar encounter with Eric, a 95-year-old medicine man who lived in a small house by the river bank. We talked a bit about my grandfather and how I wanted to honor him, and Eric told me he would help if I came back the next morning.

I arrived at Eric's the next day at 9 am and found him already starting to prepare a fire. While he placed wood into the fire pit, he began to tell me about his past.

Eric was born in the 1920s, a time when the Canadian and American governments were still very focused on "taking the native out of the natives." He said that he was taken away from his parents when he was a little boy and assigned to a boarding school. He couldn't speak any English, but every time he tried to speak in his native tongue, the teacher would hit him with a stick. After awhile, he stopped talking and kept to himself. Eventually, he learned how to speak English, but he never lost the connection to his own language and his belief in the presence of guiding spirits. He said that he began to follow the guidance of ancestors by noticing little things that seemed like unusual coincidences.

He asked me to light the fire, and once it was burning, he recited a prayer in his native language while we both threw loose tobacco into the flames. Then he burned some cedar and sage and blessed the north, east, south, and west. I felt goosebumps cover my entire body when he concluded by saying: "You are now reconnected with your ancestors. They are right now watching you. They are there when you need them."

Later that day I went back to go hang out with Press. When I got to Press's house, he wasn't there, but his neighbor David was on the front lawn, pulling a wagon stacked with car batteries. With the other hand, he held two dogs on a leash. I asked him where Press was and got caught up in a conversation about Eric and the ceremony we had just performed. Eric was the oldest medicine man on the island and apparently was revered for his ceremonial knowledge. I was told that whenever a construction company discovers ancient burial sites in the area, Eric was asked to bless the remains of the ancestors.

David invited me to come to his place. I followed him into the woods. We walked in silence while he pulled his rattling cart across the dirt road. Then he paused at an intersection and looked oddly into the distance. After a while, he kept walking and said, "I often see ancestor spirits crossing the roads here."

"What do they look like?" I asked.

"They wear traditional clothes and seem to be gliding rather than walking," he said.

"What are they doing here in the forest?"

"This area has been inhabited by Native people for thousands of years," David answered. "The Canadians and Americans have never taken this land from us. It is one of the only reservations on original land. The spirits are still around here, protecting us, guiding us."

He began to tell me about a snake spirit that apparently serves as some type of guardian. He said he saw the snake once. He thought it was a pile of old truck tires, but then the snake slithered away, uncurling its 40-foot body. I wasn't quite sure what to believe, which became a recurring theme of my visit.

We finally arrived at David's place, a clearing in the forest with a few tents and an old trailer. Barking dogs were tied to the trees. In the middle of the clearing was a lawn with a couple of lawn chairs. David pointed toward two people leaning back in the seats, telling me they were his mother and brother.

I walked up to them and introduced myself, but my smile was met with cold stares that clearly contained the question, "*What are you doing here?*" I tried to summarize my story and ended by saying that I always felt connected to Native Americans but never had visited my tribe.

"Yes, we do exist," David's mother said. Her tone was hard and cheerless, and I felt uncomfortably aware of the resentment and collective trauma caused by centuries of oppression. My spiritual connection with Eric, which I felt so intensely just hours ago, was quickly fading from my mind as I considered the reality that people on this reservation had to face.

As David's mom began to tell me about the history of our tribe, I could feel the anger and desperation that had been beaten into those souls who tried to resist the colonization of America. The history of the Potawatomi seemed alive in her eyes, and I began to get a glimmer of what it must have been like to lose one's culture and spiritual connection to the manipulative power of a foreign race.

When I asked David how our ancestors lived, he pointed to his old trailer that was covered in tarps and said, "Better than this." No one laughed. He went on to explain that there used to be a thriving population on the island that knew how to live in harmony with nature and with spirit, but that Western culture had waged a war on that way of life, eventually making the island's residents dependent on a destructive and exploitative system.

Eventually, David and I walked back to Press' house, where we helped him cut some more branches. We made a fire, roasted squash and potatoes, and talked about the Potawatomi. Our conversations covered a wide array of topics, but what stuck with me was how Press had freed himself from the mental enslavement of his upbringing.

Like many young native Americans who found themselves disconnected from their culture, Press felt lost in his youth and began to use drugs. He said what saved his life was a dream. He had gone into rehab, where he dreamt that an elder from his tribe came to him and began to speak with him in his native language. He hadn't used the language since he was a little boy, because like Eric he had been forbidden to speak it in boarding school. But he understood what the elder was saying. In the dream, the elder told him not to resist, and suddenly Press found his body covered by snakes.

At first, Press told me, he was terrified, but then he surrendered to the process. The snakes began to slither into his body, moving under his skin while the elder told him that they were eating the negative energy within him. He said that the dream felt as real as a waking moment and he could really feel the snakes moving inside of him. This experience ended his drug addiction and started his quest for a spiritual connection.

As the evening went on, he pulled out a traditional hand drum and sang some songs. I put together my saxophone and joined him. I recorded our session on my phone and after we finished playing, I recorded him telling me about how he got his spirit name:

P: This one is for the wolves, calling in the wolves ... Wolves, come in and take that negative spirit with you when you leave. You see, wolves are our guardians, wolves are our protectors, in the natural world people are scared of wolves, but wolves are good, they are adopted family. Waabima-Iigan Mnidoo Inini, that means gray wolves spirit man.

W: Is that your native name?

P: No, that is the name God gave me. Well, you see, before we are born we are like stars, God picks us. Do you ever have deja vu?

W: Yeah.

P: Before your life, God shows you your path in life, he shows you that path. So in this life, you are going along this path and all of a sudden you say, 'Hey, I have already seen this.' But actually what you are seeing is your path that God showed you already. He already showed you that. You are exactly where you are supposed to be, experiencing what God chose you to do. Deja Vu, you remember that. We are given a time period, we don't know how long, he takes you in his hand, he shows you a path and you follow that. It is like a medicine wheel, there is no end to it. You continue even after your life, even after you go back to the spirit world... so God gives you a name. Like I said, Waabima-Iigan Mnidoo. There is no one else that worked with that name. Your name is Wenzl, how many people have that name?

W: There are some.

P: There is more beyond that. So when you talk to God, Great Spirit. Waabima-Iigan Mnidoo NDishnikaaz, here

I am God associating with you. If I go there and say Press Dishna Cos, How many people are there with that name? When I announce my name he knows he is talking to me.

W: How did you get your name?

P: I died and I remembered it.

W: What do you mean, you died?

Press didn't answer my question right then, but later he told me that a hockey puck hit him between the eyes and he had a near-death experience. During this conversation, he went straight into the details of his NDE:

"I remember being in a field. I came to this big field, a nice grassy field. A tree on the other side, there was a little fire burning. There was smoke coming from the fire, there was yellow smoke, red smoke, light purple, and blue. Intertwining, going up that tree. I acknowledged those colors as my spirit name. I acknowledged those colors and out behind the tree came a big wolf and looked at me."

Later he elaborated on his near-death experience, saying he felt as if he was floating above his body, watching it being brought to the hospital. "Then it went all black, pitch black, then there was this little tiny light, just the smallest purest light you could ever see… way, way far. I could look at the light coming towards me. I got scared of it. I tried to leave, the more I struggled the faster that light came towards me. I tried kicking, running, I could do nothing. Then I finally submitted to it and said, 'Ok, I am going into this light.' I stood in that light and it was so warm, so peaceful, so comforting, I didn't wanna leave, I said I wanna keep going. Something stopped me: It is not your time, you have to go back and it was like *puffff*. And I woke up back in my body in the hospital."

"So did you see the wolf after you went into the light?"

"Yeah, I was taken there."

Press went on to tell me about several other outlandish experiences. He said that he had a friend who could leave his body and visit him anywhere. He said he would see a light coming towards him and then his friend would materialize out of the light.

Curious, I asked him what his friend would say.

"We would just chat about the weather and stuff," Press responded.

I asked him if more people have the ability to leave their bodies at the reservation. He said that it used to be very common back in the day, and that as a young boy, he used to see his grandparents and their friends sitting together in their living room with closed eyes. He thought they were asleep, but then suddenly all of them would laugh together.

"They were out doing things," Press chuckled.

The next couple of days I hung out with Press and his wife, I drove his wife to the doctor and visited another reservation in hopes of finding a direct link to my grandfather. We didn't have much luck, and the one short conversation I had with a stranger who happened to share my grandfather's last name was awkward and unsatisfying.

On the drive back from a day of walking around another run-down reservation, I felt sad. It was so visible how the destructive greed and the empty ambitions of corporate America had successfully driven the culture of the natives to the brink of extinction. Then I suddenly felt the impulse to do a vision quest. I asked Press if I could do one in his backyard.

"Sure," he said, "I will put you in the bush for a night or two."

The next day we walked into the forest with a machete, a long iron stake, some blankets, and a few cloth bags that Press had packed. He didn't say much during the walk. After about a mile he pointed to a small clearing next to a swamp.

"This seems like a good spot," he said. He put down the bags and began to chop down a dozen young trees. Press wasn't in a very talkative mood. He pointed once in a while and gave me simple instructions. After we cleaned the trees he had cut, he used the iron stake to make holes in the ground, placing them along the circumference of a circle. Then he took out a bag of tobacco and sprinkled a little into each hole before we placed the young trees into them. As he did this he said "Miigwech" to each hole, which translates to "Thank you." He explained to me that the trees had died for our project, and so we needed to show them our gratitude.

We bent the trees and tied them together in the middle, then added branches that wrapped around the hut vertically. There were four layers of vertical branches that wrapped around the hut. The bottom two started and ended at a small round doorframe, also constructed out of bent branches. He explained that the first circle represented the physical body, the second the emotional body, the third the spiritual body, and the fourth the unity with the creator, which I interpreted as the transcendence of individual identity.

Press cleared a place for a fire right in front of the hut and blessed the ground with some tobacco. He took out blue, white, red, and yellow cloth and tied strips on the roof of the hut. The different colors represented north, east, south, and west. He took more of the same cloth and asked me to make tobacco pouches and tie them to four trees in each of the cardinal directions.

We lit the fire while he explained the rest of the ceremony. He took out handfuls of cedar, sweetgrass, sage, and tobacco and put them into four wooden bowls. He told me that I was not allowed to eat or drink anything over the next 24 hours, that I should stay focused on praying and offering the herbs to the fire, and that the most important thing was to make sure that the fire stayed on all night.

"Sit by the fire and pray for the world, for all countries, all tribes, and all families. If you meet any spirits, welcome them, don't have fear... Also, you might receive your spirit name."

After he left I gathered more firewood and began the ceremony. I felt very sensitive to the subtle energies in the forest. As I was praying next to the fire, night fell and I found myself surrounded by what felt like ancestor spirits. What I perceived felt very old and very angry. It felt like the missed potential of a race that had prepared to live in harmony with all beings on the planet but had been ruthlessly exterminated by self-righteous Europeans. There was a lot of anger. But I wasn't afraid of this energy because I could understand the feeling. I also felt the pain of the lost wisdom and I was completely willing to open my heart to it, to dedicate my life to the reintegration of ancient teachings.

I said my prayers, burned the herbs, and tended the fire. I managed to catch a few hours of sleep, but I kept getting woken up by Press entering my dream. In my dream, he would clap his hands and say something like, "Hey! Keep the fire burning."

Another time I was woken up by something that hovered between the dream world and my conscious awareness. It was mostly formless but I could still sense its presence.

"Why are you out here in the woods?" it asked. The question felt kind, not accusatory.

"To pray for the highest potential of humanity. To become love and to become wisdom," I responded.

The presence said: "Ok, I will ask for that." It disappeared.

Later that night, I woke up with the words *rolling thunder* repeating in my head. I noticed that the fire needed some wood. I crawled out of my hut and placed another log on the embers. I threw some tobacco and cedar into the fire and said, "Miigwech." I looked up and the fire was illuminating the trees from below.

Suddenly I felt the angry presence again — or maybe it was just a projection of my subconscious. Either way, I experienced something powerful and slightly terrifying. It must have been around 4 or 5 in the morning, and the sky was still pitch black. I tuned into this angry presence and realized that it wasn't evil, because its intentions were to find harmony with all beings. It was only that its mindset was a little unapproachable, or fear-inspiring, but its anger was towards the loss of harmony, not driven by selfish intention. It reminded me of the wrathful deities in Vajrayana Buddhism. Instead of closing my heart to this experience, I said, "Yes, I feel you, welcome."

At that moment I felt something like an electric shock travel through my body and I simultaneously heard the loudest thunder I have ever heard. The thunder was so loud, it seemed to distort the sounds of crickets and frogs around me as if I was listening to an overdriven audio file. I looked up and I could see the stars shimmering above the treetops. There were no clouds in the sky, yet it felt like I had just been hit by lightning. There was no wind, no rain, and no more thunder.

I crawled back down into my hut, baffled by this experience. Had this thunderstorm occurred within my own mind? The forest was silent, not engaging my question in the slightest way.

The next morning Press and his wife picked me up. He handed me a mason jar filled with cool water. I opened it eagerly and never drank so gratefully, although the water smelled a bit stale. We packed up the blankets and the rest of the things we had brought with us.

When we arrived back at Press's house, I was pleasantly surprised by the feast he had prepared. He had invited his sister and her husband. They had prepared pumpkins, potatoes, duck, corn, cornbread, beans, and salads. I am vegan, but Press told me that he made this duck just for me and that his brother hunted it. I didn't want to disappoint him, so I ate some duck as well. But before anyone ate, I had to place a little food of each dish on a piece of tree bark and bring it into the forest. Press called it an offering to the spirits that had guided me on my vision quest.

We sat in his backyard for hours talking about the experience I had and vision quests and sun dances in general. A sun dance is a ritual in which the participants dance to a drum beat nonstop for four days without food or water. Press said that people often fall over and lie unconscious on the floor only to eventually get back up and keep dancing. He said it is particularly important not to disturb them when they pass out because that is when they receive their visions.

Press and his relatives have done several sun dances. They told me that after the dancing some people cut into their chests, put hooks through their skin, and hang themselves off trees as a way to override physical pain through their devotion to the Great Spirit. The tree from which they hang themselves represents nature, and hanging from the tree signifies their willingness to give their lives to the spirit of nature. Press has done it before and claimed to have felt no pain. "When you drive your body beyond its limits, at some point your spirit takes over," Press said.

I also talked to Alice, Press's sister. She must have been in her 40s, but she looked much younger. Her skin was smooth and her eyes had a slight glow in them. She told me that she was studying medicine and was now trying to build a bridge between Native medicine and Western medicine.

Alice told me that she interviews people on the island to uncover ancient medical practices. Then we started to talk about the legendary giant snake that supposedly lives on the island. Joe, Alice's husband said that when he was a kid, he had a friend who ran away from boarding school and tried to swim across the river. But the currents swept him under and he saw the surface far above his head. He feared he might die, but suddenly he felt something wrap around him and realized it was the giant snake. The snake brought him back to the surface and helped him cross the river.

I asked Joe if he really believed that story. "Every time my friend tells this story he tears up," he responded. "I believe he is telling the truth."

Whether there really is a giant snake on this island that is unrecognized by Western science, I don't know. But what I learned is that in the hearts of the people on this island lives a magical world in which spirits, animals, and plants are seen as guides and guardians. I opened myself up to this world and I hope that one day a more integrative worldview will make it normal and accepted to experience a connection to the great spirit. But until then, this kind of spiritual seeking is in nearly complete conflict with mainstream culture.

CHAPTER 17:

Barriers

Growing a connection to one's essence despite cultural resistance

Western culture has evolved many belief systems and social norms that keep us from the direct experience of our true nature and the spiritual reality of our soul.

From birth, we find ourselves thrown into a culture that places more importance on external reality than on internal reality. Europeans and their ancestors haven't managed to balance their greed and hunger for power with self-transcendence, and their culture has taken over the planet. As a result, everyone has to deal with a collective mindset that denies humanity's essence and the spiritual connection we crave.

When you look a newborn into the eyes, you can feel the baby's essence. She hasn't yet gone through years of psychological conditioning from parents and teachers. A newborn still knows her own essence. She might not be able to talk about it, but since she hasn't yet filled her mind with thoughts, her essence is all she has.

Gradually a child learns to focus more and more on the external reality and the concepts and labels previous generations have developed to make sense of it. The child gets introduced to toys, video games, candy, or whatever the parents believe their child should have.

It seems that parents don't like it when children focus on their internal world.

I remember that when I was about 3 years old, I still had a strong connection to the deeper aspect of myself. I used to look into the distance and tune out all the psychological noise of my social environment. I would remain in a thoughtless state, disengaged from the physical world. This frightened my babysitter. She would clap her hands and say: "Hey, hey, don't float away."

But my mother understood. When other people asked her what was going on with me, she used to say: "He is thinking deeply, don't disturb him."

Recently I had a short encounter with a little boy at an airport. His father was carrying him through security. The boy stared at a man next to him without blinking an eye, and his gaze reminded me of the eyes of a guru. After a couple of seconds, the father began to shake his child, saying: "Hey, snap out of it." But what he really meant by that was: "Hey, come back into the reality of psychological noise."

When a person steps out of the fast-paced, noisy world of superficial interactions we share, it can frighten or upset the people around them. If someone appears to enter another state of consciousness, it suggests that there is another way of being. This is in conflict with the beliefs we have inherited.

We have separated ourselves from our essence and then continued to tell ourselves that that's just how life is. We probably never managed to fully extinguish the little voice in us that always knows that life can be much more, but because we have continuously ignored it, it might have grown into anger or anxiety. Maybe we've started taking medication to suppress these symptoms. Maybe we've managed to be perceived as normal by others who are also focused on the superficial surface reality.

If we see someone who hasn't taken that path or isn't yet in agreement with our conditioning, we feel strongly that they are wrong and should be corrected. This is a way for us to stay out of the cognitive dissonance that we would have to endure if we were to consider a different possibility. For many people, it would be legitimately terrifying to consider the possibility that human identity is a game of multidimensional beings. They would have to open the dusty boxes into which they have neatly tucked away their fears and once again revisit the vulnerability of childhood — the fear of the unknown.

As children, we know what life is about and we are perfectly OK with the fact that this knowledge isn't rational or explainable. But as we grow up, we experience a lot of pain when we discover that what humans have created on this planet doesn't encourage our essence to develop. Instead, we become part of a mass consciousness built on self-denial. Years of denial create strong habits, which is why we must relearn how to perceive our essence.

In a way, Freud was right when he said that the oceanic feeling of complete unity is a type of regression. When we connect to the deeper aspects of ourselves, we do indeed leave behind the psychological noise that adults have come to perceive as normal. We become present with our essence just like an infant. This is why monks often seem so similar to young children.

Martin Schöne told me that during his research on brain coherence he discovered that young children often have as much brain coherence as monks, but for some reason, they lose that coherence when they grow up.

Attempting to get back to that state involves more than just simple regression. We cannot just deny the anxiety and stress of normal consciousness and pretend that we are little children again. The unhealthy psychological habits society has taught us need to be reintegrated and reorganized, just as a crystal germ doesn't get rid of the minerals around it but instead integrates them into a more synchronized pattern.

On a psychological level, this requires a lot of focus and repetitive practice. It isn't easy work, but the exercises are simple. The hard thing is to cultivate a daily and focused practice and grow this other state of being within a society that doesn't know that it exists and is built to deny it.

CHAPTER 18:

In Practice

Meditation, a simple but powerful exercise

One of my favorite meditation techniques is a very simple and ancient exercise. The instructions of the technique are like this: You focus on the fact that you are aware and stay with the sense of your own presence. When thoughts or emotions want to suck you back into the vortex of your illusionary identity, you don't resist them, but you also don't pay any attention to them. You notice that the noise is there, but you stay focused on the fact that you are aware.

Through daily practice, the sense of your own awareness starts to expand and an incredible feeling of peace and joy begins to emerge from the core of your being. Through this practice, your awareness can know itself, but it requires persistence and repetition.

This experiential knowledge is very different than rational knowledge. Through practice, you experience your awareness as something beyond the world of forms. There are no thoughts, just a deep feeling of truth and peace.

This used to be a domain only mystics would speak about, but today the idea that meditation changes one's awareness is increasingly backed up by research. There are thousands of scientific studies that show how meditation changes the nervous system, the endocrine system, your genes, and even the physical structure of the brain. A good example of this research can be found in books like *Altered Traits*, written by Dr. Richard Davidson, *Biology of Belief*, by Bruce Lipton, or *You Are The Placebo*, by Dr. Joe Dispenza. We no longer need to put faith in the words of mystics. The evidence is clear. If we want to reap the benefits of meditation, all we have to do is make it a priority and put in the effort.

CHAPTER 19:

Habit and Essence

Epigenetics, the feedback loop between our biology and attention

What we believe to be true and what we are focused on influences our body. This claim is supported by epigenetics.

Epigenetics is the study of how the environment of a cell changes its DNA. The signals a cell receives cause certain genes to be up-regulated or down-regulated. Our thoughts and feelings contribute to this process because what we think about and how we feel changes the environment of our cells. If we have stressful thoughts, our body produces cortisol, a chemical signal that tells our cells that we are in danger. If we experience stress on a regular basis, our cells modify our DNA and we create biological habits based on the assumption that we live in a permanent emergency. Our body prepares itself for fight or flight situations and has little energy for growth and repair. This is why long-term stress has such a bad effect on our health.

On the other hand, if we have thoughts and emotions based on gratitude, compassion, and love, we contribute to the health of our bodies and our personalities become more giving and caring. By cultivating positive emotions, we send neuropeptides through our bodies, which activate the receptors of our cells and modify our DNA in a way that fosters health.[22]

Neuroscientist Dr. Joe Dispenza has written several excellent books on this topic. In his 2014 work *You Are The Placebo*, he wrote about a study that demonstrated that two months of meditation can result in the activation or deactivation of more than 1,500 genes.[23] In other words, people who changed their thought patterns not only became happier — they literally became different people!

We program our bodies through habitual feeling and thinking. The way you feel now is a product of your past — the choices you made, the feelings you had, and the ideas you believed. If we carry a certain set of beliefs with us for a long time, those beliefs begin to affect us emotionally and they become part of us. It is possible to throw beliefs away when we recognize that they are false, but the habitual ways of thinking and feeling still retain a lot of momentum. We feel tempted to pick up another set of beliefs that can allow us to feel the same way.

For example, if we have gotten in the habit of victimizing ourselves or blaming others, we will try to bend even the most positive spiritual teachings so that we can use them to harm ourselves or others. Perhaps this is what happened to the teachings of Jesus Christ when they were used to create guilt or punish nonbelievers and heretics. If our habit is to punish ourselves then we will tend to identify as sinners; if our habit is to hurt others, then we will try to use the teaching to paint ourselves as superior or more enlightened.

This is very common in people who have recently discovered a spiritual teaching but don't want to give up their arrogance. Their

---

[22] Bruce Lipton, *Biology of Belief*, (Sounds True, 2005)

[23] Dr. Joe Dispenza, *You Are The Placebo*, (Hayhouse, 2014)

old way of being corrupts the teaching and they keep producing the same psychological habits. This tendency in humans can explain why religions often become corrupt.

The habits of our egos are very strong, but there is no point in adding more blame. There is also no point in becoming angry at those aspects of ourselves. They are simply there and will be there for a while. All you can do is watch them very carefully and notice how your old way of being perpetually attempts to create beliefs that allow it to keep existing. At certain times, it might even strike you as funny.

These old habits are only powerful when you fail to recognize them as thoughts and beliefs that are separate from your true essence. If you believe them, you are hooked, and your energy will perpetuate a psychological drive that is telling you that you are worthless, superior, or any other story that places your illusionary identity at the center of your experience. If you don't notice this mechanism, you follow a belief-creating drive that grows into massive networks of thoughts and feelings. It goes on to affect your endocrine system, and your body gets flooded with chemicals which make the thoughts seem even more inescapably true.

But after going through this over and over again, we can gradually learn to recognize these damaging habits. *Ah, here it is again, deep breath, feel the air come in, feel the air go out, I am here, I am aware...* Gradually thoughts and emotions fade to the background and the feeling of your essence becomes more important than the stories you used to tell yourself.

CHAPTER 20:

In Practice (II)

Knowing awareness beyond thought

We can learn to know our essence as a feeling, or perhaps as the space within which our feelings occur. The word "space" might also not be very accurate, since space has evolved within awareness. We can only speak metaphorically about awareness. It isn't possible to capture it with words or concepts, because those too have been created by awareness. So it isn't inappropriate to use extremely epic language to describe this eternal, bursting, effervescent, all-pervasive knowing. I don't think that there are words that come close to capturing the kind of awe awareness deserves.

Humans often place reverence outside of themselves, but awareness is within ourselves. We have not only become obsessed with the external world, but we also looked outside for an object to worship. We have created the image of God for that purpose; we wanted to worship something outside of ourselves. We liked the feeling of worship because it brought us closer to the love that can flow through all of us. It has been rare, though, that people have dared to say: *the love in my heart is my God, and that is the thing I am worshipping.*

Meditation brings you to a non-religious place of worship. Dogmas are just thoughts. You cannot be of any particular religion when you have no thoughts, yet it is in the thoughtless state that we find the most religious experiences. I suppose it depends on your definition of religion, but even that discussion becomes meaningless when a volcano of divine bliss is erupting in your heart. The labels we choose to give these types of experiences are secondary; what is more important is a method that can produce them.

Fortunately, such a method exists. I don't think I've ever met anyone who has spent a thousand hours meditating who hasn't experienced the divine aspect of awareness. But a lot of people aren't willing to put in the effort, which is why it still isn't normal to know God within yourself.

Giving up and letting your psychological habits drag you along is the easiest path to get on, but it's also the hardest path to *be* on. The path of actively investing your time into your self-actualization is much easier, in the sense that you don't have to spend your life in misery.

You can leverage the power of your habit-forming psychology until you are being carried by the efforts of your past self. The cyclical relationship between your feelings and thoughts doesn't have to be a downward spiral. You can also feel happy because you are thinking about the fact that you have just made yourself happy. If you are focused enough, you can escalate this feedback loop until your smile stretches beyond our galaxy. If you think that synthetic drugs can make you high, you should experience what a hundred hours of silent meditation can do. Your awareness already knows how to work your psychology; all you have to do is step out of the way and let the power of your awareness carry you.

If all this is beginning to sound a bit overexcited and idealistic, it goes back to the difficulty of writing about something that is so beyond words, particularly in a culture where superlatives like *awesome* and *incredible* have been massively diluted through overuse by corporate, political and religious organizations pushing agendas.

As wonderful as freeing yourself is, it is also a lot of work, especially in the beginning when you still need to break out of the cycles your previous thoughts and emotions have created. Later, your effort starts to carry you on its own, because your habits have become beneficial. After a while, your psyche becomes a generator of inspiring thoughts and positive emotions, but before your psyche is restructured, you have to be very diligent and aware so that you do not get pulled into the momentum of your psychological needs.

Psychological needs are what Buddhists call attachments. We get attached to what we desire or to what we think protects us from what we fear and while we act on these needs or attachments, we get pulled into the illusion and into suffering. We fail to see that the peace we actually want is beyond the gratification of psychological needs.

Buddhists have recognized that changing the external world doesn't eradicate suffering. (In extremely simple and immediate cases, of course, it does. For example, when you are actively being burned by a hot iron, removing the iron does in fact eradicate suffering.) But the Buddhist concept of attachment is more applicable to psychological attachments. For example, many people are controlled by the desire to be wealthy, powerful or beautiful and the related fear of being poor, weak or ugly. Yet it only takes a quick look at any celebrity gossip rag to see that being wealthy, powerful or beautiful is no guarantee of avoiding suffering or of being emotionally fulfilled.

Instead of chasing social status, Buddhist monks used to beg for their food and direct all of their focus toward the inward journey. They used meditation as a tool to detach themselves from their thoughts and emotions and learned to watch them as passing phenomena.

Suddenly the desire to be wealthy wasn't anything else but a passing feeling. So instead of acting on psychological needs and changing the outside world, they practiced experiencing emotions without following them.

These monks considered happiness to arise naturally when the true nature of a person is freed from the needs of the human psyche. Through disciplined practice, one can learn to observe thoughts and emotions without identifying with them.

There are many different kinds of meditation techniques, but most of them have a repetitive object of focus. For example, during Anapana meditation, one is instructed to pay attention to the sensations around the nostrils as the breath enters and leaves the body. When you get distracted by a thought, you simply return your attention back to the breath.

In the beginning, it seems impossible to keep your attention on the breath, and every second is interrupted by thoughts. With patience, diligence, and hundreds of hours of practice, though, you can gradually begin to experience awareness free from thoughts. Slowly a blissful and quiet place starts to grow within yourself. Buddhists refer to this state as *zen* or *enlightenment.*

Enlightenment is a big word and a lot of people put it on a huge pedestal, but the place I am trying to describe is a humble and quiet place of thoughtless knowing. At first, this place isn't permanent, but you can continuously visit it through diligent practice. Like a plant, it grows and carries fruits. The fruits are your liberation.

CHAPTER 21:

Into the Unknown

Letting go of rational knowing and trusting the unknown

The process of transforming ourselves can be scary because we need to suspend our thoughts, beliefs, and feelings to reinvent ourselves. This can feel like we are losing control or as if we are falling into the unknown, and in a sense we are.

In Western culture, we spend our lives focused on intellectual knowledge. In school, we learn how to memorize facts, and we are graded based on our ability to calculate, read, and write. The current system doesn't teach us how to exist on a level beyond the intellect. Most people have no idea that a thoughtless state of being even exists. When you accidentally discover it, you might feel like a rock climber on a sandstone cliff, with everything you try to hold on to falling apart. Once you have encountered the mystical aspect of thoughtless existence, the rational ideas of materialistic civilization can no longer support your identity.

The mental models we inherit are practical for the purpose of becoming a functioning member of society, but they are not practical when we try to make sense of awareness beyond the physical dimension.

In your course of your life, you may suddenly become aware of a large aspect of yourself which your rational mind has never previously noticed. In this case, you become the rock climber on the sandstone cliff. The stone breaks and you fall into the unknown, but the climber who falls over and over again eventually starts to fly, because he realizes that he no longer needs to hold on to anything. The larger reality is surprisingly safe. Although it might seem terrifying, that's just the fear of the unknown, screaming like a frustrated accountant. *The numbers don't make sense. The columns don't add up!* That fear wants to put everything into a box and put a neat label on it. But in this case, the only label you can put on it is **The Unknown**.

CHAPTER 22:

Attachment (I)

The attachment to rational understating, mystical experiences, and spiritual growth

It is common to get attached to a specific method, or a specific state of being, which makes us unable to see things or experience things that are not accessible through that method or state of being.

Intellectuals often find themselves in this predicament. They are so attached to reason and logic that they can't see or experience anything beyond the subject-object relationship. Once you encounter the interconnected nature of the universe, that kind of attachment can be deeply confusing, because there can't be any reason or logic if the subject is one with the object. There needs to be a separation to analyze the object and define its behavior in terms of a causal relationship. In other words, to recognize that one thing leads to another, you need to be separate from the process you are observing. Someone who is too attached to reason and logic fails to recognize that there is a felt reality inaccessible and unexplainable through the subject-object relationship. This is why mysticism is often looked down upon by scientists.

But mystical experiences and logical analysis can exist in harmony and even complement each other if you're able to let go of the attachment to methods and values and play freely with separateness and wholeness. You can dip in and out of the dualistic nature of the subject-object relationship. Go and be the thing, then go back and translate your experience into logic, if that is what you want to do. If you are not attached, reason and logic are very useful tools.

The opposite of attachment to reason and logic is the attachment to mystical experiences. Although it is at the other end of the spectrum, it can be equally limiting. Perhaps drug addiction is a manifestation of the attachment to mystical experiences, because drug addicts use drugs to obtain the peace of non-dual existence.

The desire to feel at peace with everything is necessary for outgrowing our limiting thought patterns, and it's healthy if it motivates us to better ourselves. But it can also get in the way of our growth if we try to find peace by taking drugs, or if our ego clings to enlightenment.

When our ego clings to enlightenment, we manifest something I like to call spiritual ego. Many people who consciously try to extricate themselves from their attachments find themselves trapped in the desire to become enlightened. It's a paradox: they become attached to non-attachment and begin to take undue pride in having traveled such a path.

A couple of years ago, the universe showed me what the results of such spiritual pride can be. My band, Moon Hooch, had some shows in England, and I decided to fly to Berlin a couple of days earlier to visit my brother and run my brain waves through Martin Schöne's Brain Avatar. (I wrote about the Brain Avatar in Chapter 3 — it's the machine that creates a resonance image based on a person's brain waves. The more you meditate or the more love you feel, the more coherent the resonance image becomes.)

Before I met up with Martin Schöne, I was so excited that I thought about it every day. I told my bandmate Mike about it, and he decided to come with me. I asked Schöne if he could measure both our brains. He said that it would be fine, but that he could only display one resonance image live. He had the idea to have Mike and I interact while he recorded Mike's brain waves and projected mine in real time onto the wall. He was going to run the recording of Mike's brain waves later through the Brain Avatar and then cut together a video that shows how our interactions affected the images of our brain waves.

When we got there, Martin carefully attached the electrodes to our heads. When he was done, it turned out that he'd accidentally switched the wires, so Mike's brain waves were projected onto the wall instead of mine. Martin was surprised by what he saw. Mike's brain seemed to be highly synchronized, as the resonance image was very beautiful and symmetric.

Meanwhile I started to feel really bad. I had been looking forward to this moment for months, and now I could not see my resonance image? I was hoping that I was the one producing a beautiful image. I was attached to the idea of seeing a tangible representation of how evolved my awareness was. I could not let go of my spiritual ego, and all sorts of negative emotions and thoughts started to cloud my awareness while Martin recorded my brain's electrical impulses. After the session, he connected the recording of my brainwaves to the Avatar.

It looked horrible! There was no synchronization, no beautiful symmetry. The resonance image showed the negative emotions I experienced, not the peaceful state I thought I could generate.

This was a hard lesson, but I am still grateful to have manifested my spiritual ego in such a clear way. I immediately understood the message: my attachment to enlightenment was getting in the way. I wanted to be able to look at my brain waves in real time; I wanted to see a visual illustration that my awareness was highly evolved.

The opposite of what I wanted happened, specifically because I was attached to what I wanted.

I am still grateful that my desire to meet Martin Schöne folded onto itself and beautifully demonstrated what spiritual ego is. If you expect results on the spiritual path, you will have to deal with spiritual ego. You will become proud of how enlightened you are, or jealous of how enlightened someone else is. Saying it so blatantly is almost embarrassing, but it is true. Many people on the path have to cross that bridge at some point.

What helps me deal with my spiritual ego is the realization that the growth of my soul is really not my ego's business. When egoistic thoughts pass through my mind I say, "Hello, Ego." This is usually enough to detach from the ego and go back to a fluid awareness of the moment.

Moving into a more fluid awareness is an intricate balancing act in which we gradually give up the part of us that wants specific outcomes. We learn that desires and methods are useful to an extent, but at the same point, we realize that our idea of what should be and how it should be understood doesn't need to have such hard edges.

CHAPTER 23:

Attachment (II)

The growth that can result from attachment

One thing we haven't explored yet is the question of *why*. Why are souls going through this process of forming habits and attachments? I believe we are getting attached to habits in order to develop our will. Every habit is an automated will response. For example, if danger approaches, you want to run away. If you like something, you want to keep it. If you are jealous, you want to keep potential mates away from your partner, and so on. Being attached creates the human melodrama, and identifying with problems here on Earth allows consciousness to gather experience and develop its will. The psychological patterns that some call attachments are the structures from which we derive the meaning of our lives. Through the struggle with these structures, we progress on our path.

I believe that we get attached and suffer because our awareness is not yet evolved enough to balance the complexity of our relationships: the relationships between our nerve cells, the relationships between our organs, the relationships within our psyche, and everything else we interact with. Our human reality is brought into existence by the relationships of complex systems and when these systems are out of balance we suffer. Through suffering, our souls learn to create harmony within the complexity of our reality. Our internal and external relationships reflect the harmony we have created, or lack of it, and through the physical representation of our human reality our awareness gets to know itself.

When there is harmony among all the components that constitute our reality, we are healthy and happy. Our essence is already love-consciousness, and it knows complete harmony, which is why we suffer when we create the opposite. The suffering that results from attachment is really the urge to grow into another level of complexity.

For example, being attached to a judgmental attitude does not consider much complexity. We look at problems, simplify them by generalizing, and then attribute faults without much analysis. Many people have this attitude when people with a different worldview challenge them, but this attitude inevitably leads to suffering. Maybe the suffering is experienced in the form of physical confrontations, or depression and anger. In whatever form it comes, the soul tends to find a way to say: "Hey, it's time to get over this attitude." And this really means: "Hey, let's grow into another level of complexity, or let's balance another layer of relationships."

Our soul is a highly harmonious information field that attempts to organize the information that has evolved in this physical reality simulation. Conflict or suffering always comes into one's life when habits or instincts are not aligned with the love consciousness at the core of our being.

Difficulties are entirely natural and necessary. Without experiencing disharmony, we cannot expand the harmonious energy of our core. But we don't need to make it more difficult than it is. Our culture has somehow developed a feedback loop where difficulties are covered by shame, which makes the difficulties more difficult.

We think we need to hide our problems, which prevents us from addressing them with a lighthearted attitude. However, we need a lighthearted attitude to address our problems. If we hate them or try to push them away, it only makes it more difficult. This is why Buddhists speak about equanimity.

Equanimity is an indifferent attitude towards all experience. Although our issues seem very personal in the moment, they are really just the residue of evolution. We experience them as issues because the love that exists at our core does not resonate with the fear-based behavior mechanisms that we might have accumulated along the way of our evolutionary reincarnation process. The seed within us is unhappy when beliefs and instincts justify each other and prevent us from expanding into the loving complexity that we can be.

So the friction that we experience during attachment and drama is a good thing; it is the breaking of old habits and the expansion into higher complexity and more coherence. Even if it feels terrible in the moment, the deepest and darkest experiences provide us with the most important lessons and the greatest opportunities for growth. Once we have dealt with a particular attachment, it becomes part of our skill set. We follow a tendency until it brings us unhappiness, and then we begin to move towards balance. By attaching and detaching, we explore a certain set of relationships. For example, when you are attached to wealth, you begin to explore the relationships that revolve around power, and eventually you will find your place between greed and generosity. But to really know generosity, you first had to experience the pain greed causes and then detach from your selfish habits.

Attachment is a necessary aspect of integrating the behavioral mechanisms that have evolved here on Earth.

CHAPTER 24:

Evolving System(s)

The benefit of curiosity

If the purpose of our existence is to work on our consciousness, why is that not made clear to us? Why are there no flying yogis hovering over Manhattan demonstrating the power of meditation? Why doesn't the larger consciousness system, if it exists, manifest huge signs next to highways that say **HEY YOU! Evolve Your Goddamn Soul**!

It could be that everything is the way it is because the larger consciousness system uses evolution as a way to program virtual realities. If everything was designed in a static way and all souls were told by the larger consciousness system how to grow their consciousness, the results would be extremely predictable and the diversity the human struggle produces would not exist. On the other hand, if you put together a rule set of physical laws and the experiential data of some type of being, then let that system evolve for a couple of billion years, who knows what you might get!

Physicist and computer scientist Steven Wolfram has illustrated this idea with his Cellular Automata project. This simulation is based on a grid which has the potential to display white and black or colored pixels. The pattern of the pixels or cells is determined by a simple relational equation that specifies how a given pixel relates to the pixels around it. If you search for Cellular Automata online, you can look at the highly complex images that these simple equations produce. (I used this type of simulation to produce the image on this book's cover.)

According to Wolfram, the results of these equations cannot be predicted; you can only discover them when you run the simulation. He posits that the universe could also have evolved based on the iteration of simple equations which produce infinitely complex and never-ending patterns.

If we are also in a procedural simulation, a simulation that gains in complexity through an iterative but unpredictable process, then it would make sense that the larger consciousness system doesn't specify how we should make sense of the world, because our world view is part of the evolving simulation.

Natural selection and random mutation produce unpredictable data. Evolution creates new organisms and new ways of making sense of existence. Human perception and human feelings are a result of this process. From this vantage point it would make sense that the larger consciousness system wants us to figure everything out for ourselves because if we were told how exactly to make sense of our human existence, it would interfere with the evolution of the system. We create new information by trying to figure something out on our own. If an authority just told us how things are, there would not be much room for creative exploration.

We can only accumulate new knowledge when we dare to question everything. If the larger consciousness system presented us with absolute truths, we would evolve only in the realm of these truths. But falling into our delusions and finding our own way out is what creates unique perspectives and these unique perspectives are what make the growth of our information system so complex and interesting.

Things are the way they are because a free will system wants us to grow our consciousness by figuring things out on our own. Just like cells had to figure out how to build multicellular organisms, we also have to find creative ways to collaborate and integrate diversity. No one told primitive cells how to work together; they had to develop a biological expression of the will to benefit each other and build the necessary infrastructure to perpetuate this will in physical from. Perhaps the larger consciousness system nudges evolution in subtle ways, but it does not tell us what to do. That would drastically change this reality and make the outcome very predictable.

Maybe there are some realities you can incarnate into without amnesia, and everyone knows everything, but this is not the game we play here on Earth. We purposely forget our real identities so that we can be driven by curiosity and creativity as we try to uncover the hidden truths.

It might seem cruel to let a bunch of souls with amnesia wander around clueless on a big rock that is spinning through space, but we are not left completely adrift. We are only happy if we do what we are here to do, and if we get lost, we become unhappy. Happiness is like an internal compass that automatically guides us. When we are truly happy, we are unfolding our potential. If we can't figure out on our own how to be happy and how to unfold our potential, we can always ask for nonphysical help. But when the help comes it often seems like the worst thing ever, because if we are stuck, we first need to break our experience bubble, and that often means confronting our worst fears.

At other times, help manifests itself in the form of seemingly random events. While writing this book, I have often asked for help and it often came in unexpected ways. Someone I have never met wrote me a Facebook message with a link to one of Tom Campbell's lectures. His theories encouraged me to take my experiences seriously and explore the simulation hypothesis.

In a more dramatic example, I was in an airplane, sitting next to an old woman. I was reading a book that happened to involve a woman named Barbara. I got a strange feeling that woman next to me was also named Barbara and was there to guide me. I turned towards her and asked her what her name was. Sure enough, it was Barbara. I was stunned but I tried to not to show it, and we talked a bit about her dry cleaning business in New York.

Suddenly, she reached over and started pressing buttons on the touch screen in front of me. "Here," she said, "you need to watch this movie." She had pulled up *Heaven Is for Real.* I watched the movie, which is how I got introduced to the idea of near-death experiences.

Many such events happened to me that slowly allowed me to make sense of my experiences. I began to believe that my teachers were always around me and teaching me through the experiences I have, and that my growth process was supervised by more evolved souls. It seemed that I could even ask them to accelerate my development. Whenever I did that, I would spend a few days meditating through intense emotions. Usually, it feels good when I meditate. When I asked my teachers to speed up my growth, though, my meditation became very unpleasant. I had to confront deep fears and I found staying calm to be difficult. If the emotions got too intense, I would ask to have my development slowed down. It was a nice feeling to believe that I was not just a conscious ape in a random universe and that I'd found a way to trust in the unknown.

I've now come to believe that we are part of a consciousness system that cares about the development of our souls. At the same time, it does not want to interfere with the path we choose, since we are in a free will system. Asking for help does not contradict free will, since making that request is still a choice we make and we choose from the paths that are presented to us. Even when we know that we are on a beneficial path, we encounter many riddles and dead ends that make us curious.

The idea of growth towards our potential is somewhat illogical because what grows is the essence of our being and not the belief systems and rational ideas we have created. We can think of those like scaffolding we use to shape our essence in physical form. But belief systems are there to be outgrown. Every belief system that brings one to inner peace and benevolence is an effective tool and part of being human involves the creative process of building one's own beliefs.

Perhaps this is why there are no flying yogis hovering over Manhattan telling us to meditate, and this is why there are no self-manifesting billboards telling us to evolve our goddamn souls. With our unique beliefs or unique ways of navigating our psychological needs, we invent new ways of what it means to become love and to be self-actualized. Just like Steve Wolfram's simulations produce unpredictable patterns, we create new ways of being. The way we deal with our challenges and the solutions we find to our unique problems might even be surprising and educational to the most evolved beings in the universe. The purpose of a simulation is to create something that couldn't be imagined.

CHAPTER 25:

Dealing With the Expectations of Others

Resisting the pressure of social norms and being yourself in the midst of conflicting expectations

We are conditioned by our social environment to care about the expectations of others. When we believe that other people are thinking about us, we tend to be interested in exactly what they're thinking, and whether it's good or bad.

I can feel this psychological mechanism very strongly when I do something in public that's outside of social norms. For example, I often practice Qigong in gas stations, banks, and the like — not because these environments are particularly conducive to spiritual practice, but because I often find myself needing to adjust my energy before I can find a private location. Sometimes I go into prayer rooms in airports, but I don't mind doing my practice in public. I actually find it interesting as a way to confront my relationship with social norms.

Social norms are mostly arbitrary concepts that are loaded with the psychic energy of expectations. This is how societies keep their culture intact. But cultures aren't absolute value systems; they change over time. When you start doing Qigong while you are standing in line at the bank, you only appear to be acting strangely because you're part of a culture that expects people to be minimally expressive, often to the point of being intentionally dull. Everyone is subconsciously expected to strive towards an idea that exists outside of themselves or doesn't exist at all.

This starts with parents who expect that we behave in a way that they think is good and with a school system which molds us according to its values. As children, we're confronted with a social environment that expects us to be different than we are and does not encourage us to unfold our essence. This is the case because of the human tendency to find comfort in the known. We become shaped by a culture and then shape those around us because we have identified with that culture. But the essence of a child isn't aligned with any culture. Something completely new expresses itself through every child — if its essence is encouraged to unfold.

If we acknowledge that our experience and our values can at best be a guideline for a child's path, then we might stop trying to shape children in our own image. But until this attitude is more widely adopted, souls in this society still need to confront the expectations of our social environment.

Most people become adults before they realize that their social environment has stripped them of their individuality and has created an identity for them. The concepts of "finding yourself" and "soul searching" only exist because of Western culture's inability to recognize and encourage the essence of a child. The same is true with the struggle between ego and soul, or between small self and true self. Such concepts only exist because our environment forces us to develop a personality that isn't aligned with our true nature. We grow up and forget who we are and what our purpose might be. We let this happen to us because we react to the expectations of others.

To find one's authentic self, it is necessary to let others think what they want to think. As souls we have to confront the social norms we are born into because our society is built on many social systems that do not serve the growth of the soul. This means we might have to confront what others believe, or at least confront it within ourselves. This is often difficult because we care about others. But if we truly care about others, we do them no favors by contorting our true essences to fit in. When we do that, we are serving nothing except the fears that drive many cultural patterns.

Our culture's loss of a spiritual connection has left us looking for gratification from the material world and from other people. Even if we never consciously realize it, we expect others to do something to make us happy. Yet this expectation doesn't lead to happiness because when we get what we wanted, we either want more or want something else. As long as we try to use an external reality to make us happy, we are overlooking the fact that the path to true fulfillment is within us.

When we start discarding limiting beliefs and start paying attention to our essence, this essence grows and fulfills us from within, regardless of what others think about us. But this means accepting the discomfort that results when you start to act in a way that others might not be able to understand. This doesn't mean that you need to be inconsiderate. When you notice that something you do really bothers someone, it could be that there is a lesson for you to learn, or that their reaction is based on their own resistance to the process you are going through.

When you begin to liberate yourself, it can happen that others who are not working on transcending their conditioning try to pull you back. If you start to know yourself beyond the things that they think are possible, then you make them wonder about how they are living their life. This can be painful because people are often tied down by fears that hide behind beliefs. If you make them question their beliefs, they also have to readdress their fears. To say the least, this isn't a comfortable process to go through. When that process isn't treated with awareness, it can lead to mean-spirited motivations.

On the other hand, the belief that others are trying to hold you down can also be very limiting. If you always explain the confrontations you have with others with the idea that they are trying to hold you back, you could be missing a lesson. Perhaps the discovery of the spiritual path is making you think that you are better than others, or that you know how others should live their life. Maybe you think that if you point out their false beliefs, they will also start to work on unfolding their essence. This is not how it works. You can only liberate yourself. When you do that, you help make it possible for others to do the same, because you no longer perpetuate the limiting social constructs of our society.

CHAPTER 26:

Dealing With Intense Emotions

Overriding the fight or flight response with focused awareness

It is extremely challenging to stop the circle of negative thoughts and negative emotions. They perpetuate each other and bring our whole being into a perceived state of emergency. It is difficult for our awareness to interact with a body that is perpetually executing the fight or flight response.

The fight or flight response is designed to engage instincts and shut off analytical thinking. When you are stressed out, your body restricts the blood flow to your frontal lobe so that it has more energy to mobilize the systems that are related to that fight or flight response. It is difficult to calm down a body that has prepared itself for an emergency, because thoughts are moving so fast and the emotions seem so urgent. It takes a lot of practice to be able to interrupt this circle and to override the sense of urgency our thoughts and emotions can create.

When it is focused, though, our essence has the power to override chemical processes within the body. Through the practice of meditation or any other mind-body practice, you gradually get the sense that your focus is something like a laser beam that you can direct towards processes in your body. When you use your focus in combination with intentional breathing, you can clear up feelings of anxiety within minutes, so long as you manage not to perpetuate the thoughts that are associated with the feelings. This is what you learn in meditation. Once you are used to letting thoughts pass, you can observe their energy without believing in them. If you observe a panic attack in terms of buzzing and tingling pressure waves, you allow your cells to make their own decisions, because you are no longer prescribing any central meaning to the panic attack and there is no longer an image or association in your mind which you resist.

Ordinarily, one's consciousness would be affected during the experience of stress hormones, but focused attention gives you the power to remove yourself from the experience. Stress is still an intense experience, but you can allow the chemicals to pass through without creating more worrisome thoughts. Your experience becomes decentralized in the sense that you, the representative of trillions of cells, don't have an opinion about your experience. By suspending your beliefs and thoughts you leverage the power of your body to balance itself. This of course isn't easy, but it's achievable through the practice of meditation.

Studies of the placebo effect have shown that what we believe has a chemical effect on our body. This effect is usually observed by the medical industry during the process of testing new drugs. When a new drug is tested, researchers give it to a group of people and note the results; those results are compared to those of a second group of people who believed they had received the medication but had only received a sugar pill. Test subjects who receive only a sugar pill often show physical and psychological effects, as if real medication had been administered. Tests like these forced the medical industry to acknowledge that belief alone can affect a patient's neurochemistry.

Dr. Joe Dispenza writes in his book *You Are The Placebo* about a case study in which the medical facility in a World War II military camp ran out of morphine. Knowing about the power of the placebo effect and not wanting their injured to suffer, the doctors decided to give the wounded soldiers fake injections while telling them that they were getting morphine. The soldier's expectations were so great that their pain instantly disappeared. Based on this and similar case studies, Dispenza argues that the human brain is effectively its own pharmacy. What we believe sets in motion complex processes that produce the chemicals that deliver the physical results we expect.

This isn't only true in relation to drugs. Any belief which is related to the nature of our experience affects the body on a chemical level. Your opinion about anything in your reality affects your biochemistry. You are constantly creating chemical realities for your cells.

Most of us have experienced the sensation of hearing terrible, unexpected news and suddenly feeling a rush of adrenaline. That's your belief in the information you received triggering the release of a chemical. The key element here is that it doesn't make a difference if the information you received was true or not. Even if someone pranks you, it fires up your adrenal glands. Your belief in the information is the deciding factor.

Our beliefs have the power to medicate our body, and that's the reason why there is such a thing as the placebo effect. But what happens if we do not believe anything? If we just pay attention to how our bodies are feeling? What happens to our bodies' capacity to generate chemical realities when we no longer tell the body what is true and what isn't?

My hypothesis is that we empower our cells when we suspend our beliefs. In this state, we are no longer telling them what emotional states to create; instead, we are "listening" to what they are experiencing. Our cells' language is chemical, and our understanding of their chemical language is our experience of emotions. So when we quiet our minds and disregard our thoughts, we can use our focus to scan each area of our bodies. This meditation technique is called Vipassana meditation.

During Vipassana meditation, we gift our body our nonjudgmental and undivided attention without creating emotions by an imagined future or past. When you go very deep with this technique, you realize that your body remembers traumatic emotions and begins to optimize itself. You might not understand the emotions this process dredges up, and you might feel a need to justify them with reasons in your environment. But during a silent meditation retreat, it is hard to blame anyone, since you don't have any interactions. So you get to observe the mechanism between thoughts and feeling, separate from external influences. The body brings up strong feelings which are often unpleasant, and the mind then tries to find a reason for these feelings. But ultimately they are just emotional habits, or communication networks between cells.

The kind of data exchange that is happening on the level of cells is massive and far beyond the rational mind's ability to understand, yet we feel the need to integrate these networks in rational understanding, so we try to find reasons in the environment. In ordinary circumstances, we find some sort of balance between our need for beliefs and our experience. But in a meditation retreat, that balance becomes lopsided as you are invited to see the mechanism of your psyche functioning in isolation.

It is difficult for control freaks (including me!) to surrender to a cellular reality which has no rational meaning. Often the mind tries to approximate the processes that occur on a cellular level. But this just creates new cellular realities. I feel things, then form an opinion about them, and in the process change the feeling, which then requires a new explanation. This is a feedback loop that can become unhitched during intense meditation. If you find a fearful belief that corresponds to your fearful feeling, the feeling becomes more intense and the belief seems more true. The same can be true for positive thought and positive belief. You can spiral up into complete ecstasy. But as you sit for hours on end, you start to see through that mechanism, and you no longer believe the realities you are generating through the circle of feelings and beliefs. So you just watch them spiral up and down and breathe deeply while you let your cells figure out their own reality.

It is amazing what cells can do if you are willing to step out of the way and lend them your focused attention without injecting your opinions and beliefs. Those beliefs and opinions will keep manifesting themselves as thoughts, but you do not need to follow them into the realities they want to explore. You can see them like ripples in a lake. Negative thoughts and emotions are stronger ripples, but they too pass and there is no need to engage them, as your body knows how to handle its own reality if you remain still and observant.

Harvard scientist Richard Davidson has spent his life researching the physiological effects of meditation. He has analyzed hundreds of detailed studies that explored the psychic and emotional effects of meditation. The intersection between science and spirituality is often muddied by pseudoscience or scientists who accept questionable data just because it supports their hypothesis, but Davidson sets himself apart from many researchers through his rigorous approach to the field of meditation research.

In Davidson's book *Altered Traits,* he says that the findings of meditation studies are often imprecise because the researchers fail to recognize that the environment within which the case studies are conducted also has an effect on the neurochemistry of the participants. To isolate the effect of meditation, he used control groups who were in the same environment but did not meditate.

One of the studies he writes about measured the effect Vipassana meditation has on the attentional blink. The attentional blink occurs when a subject is asked to recognize a few symbols in a string of many symbols. After a person recognizes a symbol, the brain is busy processing that symbol; during that brief moment, the brain doesn't notice any other symbol that follows. Neuroscientists thought that this was a hardwired condition. However, Richard Davidson claims that studies on Vipassana meditation showed that a three-month retreat can decrease the attentional blink by 20 percent!

Another study revealed that long-term meditation creates more connections between the frontal cortex and the amygdala. The amygdala is the part of the brain that is responsible for the fight or flight mechanism. Long-term meditators learn to control this part of the brain and reduce the amount of stress they experience. The brain grows new synaptic connections and optimizes itself. According to Davidson, even as little as eight hours of meditation creates measurable brain changes.

This might seem strange. You sit down, think about nothing, and your brain rewires itself? How is that possible? What type of intelligence is responsible for these changes and why do they occur when we learn to stop thinking?

I believe that meditation optimizes the brain and the body because our cells are trillions of intelligent beings that respond to a central voice, which are our thoughts and our beliefs. This central voice is like God to them, because what we think and believe changes their entire reality. It doesn't really matter if we're stressed out due to an imaginary scenario or if we're stressed out due to being chased by an actual lion. Our cells get affected by the same chemical reality.

When we stop thinking and start feeling what our body is feeling, we become like a dictator who renounces his throne and gives all of his power to the people.

Fear-based societies with central power structures never build optimal environments for their citizens. The centralized nature of a dictatorship simply doesn't allow for the flexibility and awareness needed to solve the diverse problems of its citizens. In the same way, a body dominated by a few overactive neural networks will always be a stressed body. The ego personality is a network of brain cells that is very small compared to the rest of our neural networks.

When you learn to suspend your thoughts and feel your body, you empower the "oppressed" cells that never receive any attention. During meditation, you take the attention away from the default mode network (the neural correlate of the ego) and hand it over to the rest of your body. When you focus on a part of your body, you increase cellular activity in that area. It is no surprise that this process can bring up painful emotions.

Imagine what would happen if the people in power suddenly became acutely, viscerally aware of all the suffering that occurs in the slums of Asia, Africa, and South America. Imagine what would happen if the people in power stood still and felt the pain of oppressed workers who have no choice but to work underpaid jobs under terrible circumstances. If this kind of emotional suffering was truly perceived by those on top, the world would either become a very different place, or the people in power would have to try even harder to shield themselves from the suffering their agenda creates.

This can also happen during meditation. You might suddenly become aware of emotional pain that you haven't felt since your childhood. At that point, you either open your heart and surrender, or you tense up and try to protect yourself from that pain.

It takes a lot of courage to surrender in the face of painful emotions, but we know now that our bodies are far smarter than we are. Not even the most revered neurosurgeon would be able to build a functioning network between the amygdala and frontal lobe or restructure the brain to reduce the attentional blink. But we can tap into the decentralized power of our body when we give up control and allow our cells to optimize their environment.

British philosopher Alan Watts used to say that a good leader knows how to delegate responsibility. A decentralized organization is also called a "heterarchy," which is a term that refers to a type of organization in which the lower levels affect the higher levels just like the higher levels affect the lower levels. (In a hierarchy, the top layers control the bottom layers.)

A heterarchy is superior to a hierarchy in the sense that it is more responsive to the environment and therefore can deal with more complex challenges. Imagine a world in which all citizens were given the power to build the environment they want to inhabit. The solutions humans would create if we were to adopt more decentralized organizations would far exceed the capacities of centralized governments. In the same way, our bodies can optimize themselves if we lend our cells our neutral awareness.

To restructure your cells' relationships, you have to listen to them. You don't speak their language — or, at least, you cannot translate their signals into rational meanings. Maybe you can find explanations for why you are feeling a certain way, or what caused your emotions, but you don't understand the deeper meaning. You don't understand the changes your cells would like to make to your nervous system and so on. To allow your cells to optimize your body, you have to be willing to be present with your body and every feeling that might arise without distracting yourself with thoughts.

Maybe your cells would be thriving more if your amygdala was connected to your frontal cortex and if your anxious thoughts weren't firing up your entire nervous system. A nervous system that releases adrenaline while you are worrying about the future is like a government that announces a national emergency because the president isn't feeling well.

How often do you actually have to face the exact future you are worried about? Most of the time, the future you end up experiencing isn't even close to what you are worried about. The cells of most people live in an environment that is far from optimal, but to allow our cells to restructure our bodies, we have to be willing to experience anything our bodies want to process.

This means that we have to shift our attitude toward our emotions. Instead of trying to find solutions in the physical reality, we have to grant ourselves the time to experience emotions without the need to find rational solutions.

Our focused awareness itself is the solution for our cells. If we stay fully present with the sensations within our bodies, they will transform intense emotions by optimizing neural pathways.

Being patient with whatever state you are in now is critically important. The saying "neurons that fire together wire together" is well-known among neuroscientists. For us, it means that the ways we feel and think become habits. If we have thought and felt a certain way for many years, we cannot expect to break such habits immediately. In fact, such expectations only create more stress and tension. It is more productive to give up the will to change ourselves and allow all our stress and negative emotions to exist exactly as they are. The paradox is that accepting our emotions as they are actually transforms them. Stress needs resistance to perpetuate itself, even if that resistance is the resistance to the feeling itself. So we change ourselves by giving ourselves permission to be exactly how we are.

It's important to note here that there is a difference between allowing ourselves to be how we are and falling back into unconscious habits. When we allow ourselves to have negative emotions, we remain aware of them and we remain aware that they try to justify themselves with whatever belief system seems appropriate. We watch this process indifferently, giving it permission to exist as it is, but we don't go into it and reinforce the negative emotions with whatever belief systems they try to fabricate.

But this only works if we do it sincerely, if we really allow this process of feeling and thinking to be as it is. This is difficult because very negative emotions and thoughts can make us think that we are inherently bad people. If we respond to that belief, we then try to hide the process from ourselves while we develop self-hatred. If this process of self-delusion has gone on for a while, it is difficult to step out of it and allow it to be there.

What helps here is realizing that you aren't anything your thoughts and emotions have tried to make you identify with. This realization needs to become habitual. This means that you have to understand this all the time, not only in reference to the way you currently feel. It is easy to know your essence when you feel good, but it is hard to know your essence when you feel terrible and have an object of blame. It is this challenge that makes human life an educative experience. The feedback loop between thoughts and emotions has the power to claim our identity, but with the right tools, everyone has the power to rewrite their programming and allow their cells to organize themselves more optimally.

I believe that this self-organizing capacity of your body is driven by your essence. It is the field of awareness or your soul that puts your cells into a relationship with one another. If your ego and rational processing are quieted, you empower the deeper level of yourself. The emotions that come up during this process are the difference between your physical self and your nonphysical self. They are the pain that has forced your body to grow away from your soul's architectural plan. I believe that the restructuring of the body that can happen through meditation is the soul's attempt to reclaim the body and heal those aspects that have split off during traumatic life events. The mainstream medical community is well aware of the neural deformations that can result from child abuse, but we are also well aware of the body's healing capacity.

Although I cannot prove it, my own experiences have led me to believe that this healing capacity is essentially the soul's ability to create harmonious relationships and its ability and desire to be expressed in physical form. To tap into this innate power, all we have to do is quiet our minds, allow our inner wisdom to take over, and tolerate the discomfort that might arise.

Depending on how deep you want to go with this, your soul can become so active that it feels like a laser surgery. At that level, the rational frameworks begin to break down as you find yourself interacting with a transpersonal intelligence that is both you and not you. The wisdom your soul is connected to can appear to be a huge network of nonphysical beings that upgrade your body to become a physical expression of wisdom and love.

Obviously, these claims are far from being recognized by science, let alone proven by science, but I decided to include them at the end of this chapter because I myself was surprised and startled when I first found myself in that type of reality. Since then I have met other people who have described similar experiences. These days, meditation is in fashion for health and relaxation, but you can also use it go all the way to a point where you give your body and everything you are to your soul and the network of intelligence which it is part of. This means that the identity your soul has built in time and space gives itself to its essence.

CHAPTER 27:

Meditation and Pain

Finding peace by accepting pain

I write a lot about meditation because it is my favorite practice. But I think that the beneficial essence of meditation is available through any practice that sharpens your focus and your ability to direct it. Exercise like jogging, swimming or skating can be meditative if you focus on your breathing and repetitive movements. Cutting onions can be meditative if you stay focused on the sounds you hear and the sensations you feel. Meditation is a mindset and isn't dependent on a specific activity; it is the intention to be present. It helps you restructure your relationship with thoughts and the non-rational experience of the present moment.

The present moment is a direct experience, not an analytic concept. When you are in your body, the present moment is the experience of your internal and external environment. When you are present, your awareness is focused on the data stream of your senses, not on the rational interpretation of your experience.

Being fully present with our direct experience can be deeply stressful or painful. When we begin to fully experience ourselves, we begin to notice all these fears or unresolved emotional traumas picked up during our childhoods or perhaps during past lives. These psychological tensions are uncomfortable, and we may find temporary relief when we distract ourselves with the world of thoughts. But compulsive thinking doesn't spare us from the experience of our pain. Even if we successfully flee into the world of thoughts and convince ourselves that we don't have any emotional pain, we will discover our pain in a different way, because our thoughts will be driven by the pain we aren't willing to feel. Feeling pain is an integral part of physical life. We can distract ourselves with thoughts or external stimuli, or we can learn from the pain by allowing ourselves to be fully present with whatever we feel.

Although meditation ultimately elevates pain, it doesn't do so by pushing the pain away or pretending that it doesn't exist. Meditation is a method that retrains your mind to stay focused on your direct experience, whether it is painful or pleasant. During long retreats, it is common to experience excruciating pain as you sit for hours on end. But through constant practice maintaining a calm mindset, even the most unbearable pain can become a neutral signal as compassion, joy, and love take its place.

These elements also appear in other spiritual traditions. When I talked to my Native friend Press about his experience at a sun dance, he told me that the ceremony is about overcoming pain through devotion. You demonstrate that your love for the great spirit is stronger than your pain, and through this determination and endurance you experience liberation, because your focus is freed from the emotions you were trying to avoid.

Overcoming pain seems to bring a tremendous emotional reward. Unpleasant emotions or physical pain normally direct our consciousness. When we accept pain and don't mind experiencing it, we are released from the control that pain usually has over our consciousness.

One way to think about this is that your awareness is strengthened by breaking the control mechanisms of the human psyche. Normally, your essence is overridden by mechanisms that we experience as pain. When something physically hurts, most people immediately discard whatever they were thinking about and do whatever prevents the pain. But overriding this automated response can be very beneficial for the soul because suddenly we aren't acting based on physical survival. Our actions become more than the psychological inheritance of a physical creature.

The process of growing our souls is partially the process of un-identification. When our souls aren't adequately expressed, our behaviors are driven by instincts and habits. When your soul grows towards its potential, you gain the ability to experience instinctual urges or emotional urges without the need to act on them.

Learning to remain dedicated in the face of pain removes the fear of pain, which widens your spectrum of choices and strengthens your willpower. If you don't fear pain, either emotional or physical, you are generally less afraid, which allows you to make more choices based on love. If you no longer need to protect yourself from experiences you don't want, your focus shifts from self to other. I believe that this is why martial artists or other people that push themselves to extremes often have a very positive and kind mindset. They are no longer afraid of pain, which gives them a certain level of liberation.

Meditation is similar to martial arts in that way. Extreme meditation can bring forth tremendous amounts of physical and emotional pain, and after sitting through hours of misery the pain is transformed into lighthearted joy. This is what I often experience during the retreats.

I believe that confronting discomfort is a crucial part of growing your soul. Whether you choose to follow an extreme practice or whether you make the emotional pain of daily life your spiritual practice, the results might be the same. Suffering is everywhere, and we constantly have the opportunity to learn from it. It is a mindset that determines whether your soul grows, or whether it gets pacified by habit, conditioning, and instinct.

Pain is good for you as long as you know the tools of liberation. These tools are the knowledge of your own eternal nature and the knowledge that physical survival isn't the ultimate goal.

If you combine this knowledge with a practical exercise that strengthens your focus, you are equipped to deal with whatever life throws at you. If your significant other breaks up with you, you can see that as your spiritual exercise. Your chest will probably be filled with acute pain and your mind will get flooded with thoughts of desperate longing. Intense breakups can expose many painful patterns that are embedded by habit in our psyches.

When we go through a breakup, we get thrown into very inharmonious emotional territory, which is very unpleasant but provides a huge growth opportunity. The pain we experience during a breakup is created by our inability to understand eternal love. In this physical reality, things really seem to end, and it is painful to experience the illusion of separation.

By allowing yourself to feel a painful emotion, you allow your soul to integrate the energy of that emotion. Emotions are energetic patterns of some type. Perhaps you can think of them as probabilistic information fields, since they are guiding your probable choices. Emotions are a certain potential because they motivate consciousness towards specific actions. When we feel an emotion, there is a certain probable future associated with it, but on a practical level, it doesn't matter if we think about emotions as probabilistic information fields because thinking about them isn't the same as feeling them.

Negative emotions have power over our will if we don't allow ourselves to feel them. They are like programs that can run our organism, but only when we try to avoid them.

When a painful emotion enters our awareness and we refuse to tolerate its existence, then we are controlled by that emotion. Every time our soul gets in contact with that program, we run in the opposite direction, which means we lose our willpower to that emotion. But understanding this once isn't enough. To really integrate emotional pain, we have to confront emotions with patience and acceptance over and over again.

This is extremely counterintuitive, to say the least. Evolution has programmed us to take quick and definitive action in the face of threats. But at the level of complexity at which we exist now, sometimes things need to be the other way around.

Painful emotions create painful actions. We are caught in a chain reaction which Hindus refer to as karma. If we're to break it, we have to resist survival-based drives and confront pain with a calm and loving acceptance. When you stay anchored in this awareness, it can feel like a powerful beam of light that pierces the darkness and lifts every aspect of your reality to its highest potential.

The soul's growth is driven by repetition. To override a pain-based program, that pain needs to be liberated over and over again. We have to spend much time with things we don't want to feel and focus on remaining calm and accepting.

Sometimes I sit in meditation for hours, focusing on my breath while feeling an unpleasant emotion in my chest. Sometimes a specific emotional trauma takes months or years to be processed and integrated, because what you are learning isn't just coming to you on a rational level, it is affecting a deeper level of your being.

It might seem easier to repress painful emotions or drown them out with pharmaceuticals, but in the end, you are just postponing the development process. The emotion remains an information field that you don't want to process, but at some point your soul needs to process that information field to reach the next level of complexity. Your soul's potential wants to express itself in emotional territories in which it hasn't expressed itself yet; that is the nature of growth. The abilities we acquire in this territory might determine which worlds, or experiential realities, we can inhabit in the future. As Robert Monroe would put it, your level of maturation determines which belief system ring you inhabit — or if you leave form-based realities altogether.

During a meditation retreat in my bedroom I experienced communication with nonphysical beings that congratulated me on my human life and encouraged me to finish it. But they also told me that they would be excited to welcome me in their reality after I am done here.

A couple of years later I dated Julia, the girl that visited me during an out-of-body experience. Several years before I met her, she told me, she had encountered a transpersonal intelligence which she sometimes referred to as aliens or guides.

We were sitting in her living room next to her cat Thia. I was petting her cat when Julia suddenly told me that there was a message for me. She said she could see the beings, which she described as floating above her in a ring with their heads joined together. Then she said, "You will do important things here on Earth, but after you are done here they would like for you to live with them."

I laughed and thanked them. When I asked Julia where her alien friends were from, she said "Space Trone, Supernova, complex B." This didn't mean much to me, but it certainly made me think. Maybe these beings are the energy that we perceive to be an exploding star. In Tom Campbell's MBT and Donald D. Hoffman's MUI theories, a star is also organized by probability or awareness with potential. What the reality of that awareness is like is completely unimaginable to us. It could be that the energy that organizes the growth of a star is engaged in a virtual reality that we cannot perceive. Who knows what relationship that energy has with the reality we perceive?

Every time my ex-girlfriend spoke with her guides, it raised more questions than it provided answers, but it strengthened my belief that there are many more realities beyond space and time.

A famous mystic from India, Paramahansa Yogananda, once wrote about an experience he had after his guru passed away. While Paramhansa was meditating, he suddenly saw his deceased guru standing in front of him, surrounded by a bright light. He told Paramahansa that he had become a teacher between several worlds and he helps beings evolve from form-based realities into formless realities.

Robert Monroe has also talked about nonphysical experiences in which he tried to help recently deceased humans get accustomed to formless existence. As he flew with them through the belief system rings, he said, they disappeared at a certain level. He said that they would stay in a belief system ring because they emotionally agreed with the reality the souls there had created. This doesn't necessarily mean that the souls there would agree with each other — they would simply agree with a certain way of being, even if that agreement would cause them to fight with each other.

According to Monroe, the beings in a belief system have a similar level of tolerance towards emotional energy fields. When they feel the energy of fear or desire, they either act on it or have the ability to defuse that energy emotionally. The more advanced a soul is in its ability to defuse and integrate emotions, the more options that soul has, because it is no longer confined by what it tries to avoid.

Michael Buhlman writes about the same thing, but refers to the different kinds of realities as "density levels." According to him, fear-based realities are denser because a stricter ruleset is needed to facilitate collective actions. By "ruleset" he also refers to the rules that make up our physical universe.

This is all very heady stuff, but we don't actually need to understand this rationally or even believe in it to reap the benefits of emotional growth. The benefits of growth are immediately relevant to our current lives, because if we grow out of our fears, we feel better.

For some people, it might be beneficial to think of their life's troubles as challenges that they're facing to contribute to a growth process that transcends one life and one reality. Others might not need that belief, since growth is beneficial even if you only believe in one life. If you change your perspective of suffering and pain and cultivate focus and self-knowledge, it will have a real effect on the rest of your experience here on Earth regardless of whether you believe in an afterlife or not.

CHAPTER 28:

The Collective Mind

My personal experiences with the collective mind and William McDougall's experiments with group learning in rats

I got interested in the idea of the collective mind because of the subjective experiences I had during deep meditation. During one such experience, I perceived the expansion of my identity. Usually we feel that our selves end where our bodies end, but this is just one mode of perception. After several days of quieting my mind, I had an experience in which I felt that my sense of self included several layers of formless information fields that were inhabited by discarnate beings. I felt that the concept of "myself" included millions of different souls that all had different beliefs and different agendas.

During this novel state of consciousness, I began to notice different cultural agendas created by groups of beings. For example, I felt a Christian thought system that was spread across the minds of countless beings. I didn't see these beings, but I felt their relationship to each other and how their thoughts and emotions played out within a certain framework of beliefs and created patterns in relationship to different groups. I wasn't sure what the reality of this experience was, but I could observe how each thought I had affected this Christian group mind. This experience was as if I was thinking and perceiving from the vantage point of a mind that encompassed this field. I experienced the interactive dynamics of thoughts and beliefs within a transpersonal field of consciousness. The fact that this experience was in the context of a Christian belief system was particularly unexpected because I was raised as an atheist and never had anything to do with Christianity.

During meditations I had several experiences of this nature, and the idea that our thoughts and emotions contributed to a collective information field had become part of my subjective reality. But as it turns out, meditation isn't the only way to experience the effects of a collective mind. Experimental evidence for such an idea has been collected throughout the last century.

In 1920, psychologist William McDougall wanted to test the hypothesis that behavioral traits can be genetically inherited and passed on from one generation to the next. McDougall devised an experiment at Harvard to determine whether rats pass on learned behavior through genetic inheritance. Like many other scientists, he was more interested in his hypothesis than in the well-being of his lab rats. So he threw them into a tank of water.

Unsurprisingly, the rats wanted to get out, but McDougall didn't make it easy for them. He had constructed one brightly illuminated gangway and one dimly lit gangway. The dimly lit gangway was an exit, but the brightly lit gangway was pure torture: once the rats entered, they received painful electric shocks and couldn't get out. Needless to say, the rats didn't want to get electric shocks, but their fear of drowning made them panic and again and again, causing them to swim to the brightly lit gangway and suffer rather than swim to the dimly lit gangway and escape.

McDougall watched the rats get electric shocks and wrote down how many times they failed to find the exit. He wanted to see if learned behavior is passed on through genetic inheritance, so he separated the offspring from each generation and tested them independently. He recorded exactly how many times the average of each generation got electrocuted, and after eight generations he was seeing a clear downward trend. The rats were obviously learning.

To McDougall's surprise, the control group, comprised of rats from the same species that weren't offspring of the tested rats, also exhibited the same trend.[24] Somehow those rats were also learning how to avoid the shocks.

This experiment went on for 15 years. In the end, McDougall was more confused than before. How could this be possible? He had originally hypothesized that the rats in the test group could pass on learned behavior from one generation to the next through their genetic code. Since the control group wasn't related to the active group, their behavior should have stayed the same, but instead, all rats tested made fewer mistakes each generation, regardless of their genetic relationship.

Later, Wilfred Agar and his colleagues at Melbourne University took matters (and rodents) into their own hands and threw more rats into electrocuting water tanks. Agar got similar results as McDougall, except that the starting point of rat mistakes was about where the study of McDougall had stopped. Rats were learning how to get out of the water tank without getting electrocuted, "but the same tendency was also found in the untrained line."[25] Wilfred Agar had started with confidence, but after 20 years of intensive rat torture, he was defeated. The fact that rats were learning as a species independent of

---

[24] McDougall (1930).

[25] Agar, Drummond, Tiegs and Gunson (1954).

their physical contact seemed impossible, yet this was precisely what the experiments confirmed.

After several decades of experimentation, this particular type of rat torture stopped, and most scientists remained confused about the results. Not Rupert Sheldrake, though. Under Sheldrake's theory of morphogenetic fields, it would make perfect sense that rats learn from their collective mind together, regardless of their interactions with one another.

These findings raise a lot of questions about what other aspects of animal behavior could be influenced by nonphysical information fields. If rats have something like a collective consciousness through which learned information is shared, perhaps humans do too.

This idea seems to be supported by what is known as the Flynn effect. The Flynn effect is named after James Flynn, who discovered that the average score of IQ tests in America has increased 3 percent every decade since the WWII era. No one has been able to explain this increase, but these findings could be seen as evidence for an immaterial information field which human psyches interact with. The more people take the IQ test, the more data is stored in this field, which then is accessed by future generations — at least, according to Rupert Sheldrake's theories.

Although the idea of immaterial information fields is still considered fringe science, I believe that there is enough evidence to take this idea seriously.

How does this hypothesis affect us emotionally? What does it mean for us that we might be sharing all our mental and emotional information as a species? Can we even consider thoughts and emotions our own? Or are we more like transmitters and receivers in service of an evolving field? Are these fields related to learned behavior? Aren't emotions also learned behavior? Could it be that the emotions we feel are the emotions of humanity and not just our own emotions?

Decades before Rupert Sheldrake popularized the hypothesis of morphogenetic and behavioral fields, Carl Jung had proposed a similar idea. He believed that our emotions and thoughts are part of the collective unconscious — another term for the collective mind. According to Jung, the whole history of humanity, everything that has ever been felt or thought, still exists as information that our awareness can tap into.

From this perspective, each individual has to address both the personal psyche and the collective psyche. The personal psyche is a part of the collective psyche or mind that has a name and the memory of one lifetime, but the patterns of that life add and draw from a collective field. The anger you had to address during one life might be something that had been created by others before you were born. Maybe the anger that is present in your interpersonal relationships was a pattern that has perpetuated itself for thousands of years. But this doesn't just apply to negative emotions. The same could be said about ideas we have — or any other aspect of our psyche.

This view suggests that when our awareness incarnates on Earth, it connects to a collective information field, from which it draws information and to which it adds information.

During one of my meditation sessions, I experienced my psyche as layered. Each layer appeared to be associated with a particular phase in my development. Eventually, I experienced emotions that I last felt when I was just 4 years old. They were painful and uncomfortable emotions, but as I stayed present with them, they began to fall apart.

As each layer collapsed, I started to feel a new level of love and peace. As I worked through these layers, I began to feel emotions that

I didn't recognize as my own. I felt like a deep sea diver using my focus to explore the emotions of the collective psyche. I started to feel such a deep pain that I could only describe it as absolute desperation. If I had to put a story to it, I would say that these emotions were caused by people being burned at the stake, or tortured with hot irons. It felt like I had somehow accessed the collective pain that has been haunting humanity. This experience changed me because it felt as if we all share the same pain, and that the way I deal with my pain affects everyone else.

CHAPTER 29:

Belief and Emotion

Our personal relationship with the collective mind and Robert Monroe's Believe system rings

A subconscious agreement perpetuates an emotion. By "subconscious agreement," I mean a belief that a person acts on without actually being aware of. This belief relates to the nature of one's experience, and it's why we sometimes feel a certain way in certain situations even though we intellectually realize the feeling isn't justified. Usually these subconscious agreements or emotional beliefs are anchored in our past experiences. Our minds detect a situation for which we have some reference and create emotions that are intended to prepare the body for the interaction we expect. So we carry our past as emotions or emotional filters which shape our perceptions of what the future will be.

The same type of experience can mean different things depending on the beliefs our emotions are anchored in. One person might feel proud to be in a big new truck, while another might feel embarrassed, because for the first person a truck is primarily a symbol of financial success and for the other it is primarily a symbol of environmental destruction. Likewise, one person might feel good about having a one-night stand while another would feel like a sinner.

The differences in these reactions are created by our personal relationship to the collective mind, because the collective mind is the source of inherited beliefs. For example, someone who agrees with traditional Christian morals would identify with the archetype of the sinner when engaging in casual sex. Meanwhile, a person who identifies with the social norms of a promiscuous fraternity would likely brag about the many women he has slept with. If we think about this in the context of a collective mind, it changes the way we perceive our own emotions. They are unconscious agreements with the creations of our ancestors.

From the perspective of Rupert Sheldrake, the behavior of any species is influenced by shared behavioral fields. These fields are like data banks holding information about how a species has reacted to specific experiences. This is why the rats in McDougall's experiments could learn from the experiences of other rats that they weren't related to and never had physical contact with. Assuming that the same thing is true for human beings, the emotions we have could also be seen as the learned behavior of other humans. They aren't just anchored in our own past — they're anchored in the past of every human who has ever lived.

Rupert Sheldrake's hypothesis proposes that these behavioral fields evolve in relationship to certain experiences. This means that someone who believes in something our ancestors believed in enters into a relationship with the emotions and thoughts that have evolved within that belief system. Belief in Christianity, for example, would put you into a relationship with all souls that have ever shared the faith. The emotions they experienced influence you and the way you interact with these emotions influences the field and those who connect to it.

This might sound like a far-out idea, but it is useful in understanding and explaining transpersonal experiences.

Carl Jung developed the theory of the collective unconscious as an integrative tool for his own experiences and the experiences of his patients. The year before WWI broke out in 1914, Jung had increasingly unsettling dreams of a sea of blood pouring over Europe and washing thousands of corpses away. He thought he was losing his mind, but when war broke out he came to believe that he had foreseen it in his visions. Jung decided that he needed to develop a new theory about the nature of reality and the human psyche.

In his dreams, Jung often found himself communicating with mythical beings. According to Jung's writings, the information these beings gave him seemed to surpass his own knowledge, and they seemed to have a connection with humanity at large. He concluded that these beings were archetypes within the collective unconscious.

While exploring his own relationship to the collective unconscious, Jung came to believe that the main difference between himself and his patients was that he had the ability to integrate his connection to the larger reality, while some of his patients lost their focus within the currents of this deeper layer of the psyche.

Intellectual ideas can help psychological growth when they allow one to make sense of one's own experiences, so it's no surprise that many spiritual seekers or mystics come up with theories that have a concept similar to Jung's collective unconscious.

Robert Monroe used the term "belief system rings" instead of the term "collective mind." In his books *Far Journey* and *Ultimate Journey*, Monroe wrote about out-of-body experiences in which he perceived visual representations of collective information fields. I say "visual representations" because in an information-based worldview, everything that has a form is a representation of data. Perhaps it is possible to disconnect from the information stream we believe to be our physical universe and to visually experience the collective psyche.

Monroe described his "belief system rings" as layered rings surrounding our globe, with each ring inhabited by discarnate beings that agreed with a certain type of belief or emotion. Dogmas and emotions are interrelated because beliefs create emotions.

I believe this is true both in a religious context and on the level of relationships. For example, the belief that one person is better than another or that one wins and another loses is a belief that creates the emotions of competition and the archetypes of *winner* and *loser*. But to participate in this type of dynamic, one has to identify with the game and certain rules or sets of beliefs. The same is true for any other emotional dynamic; one has to agree to a certain framework in order to participate emotionally.

That is why it can seem absurd to us when people get heavily emotionally involved in a sport or a fandom that we could never take seriously. Every person has their own set of unconscious agreements and unconscious games, and the people that have agreed to similar dynamics usually flock together to play out the dynamics of their beliefs. Depending on what these subconscious games are, these dynamics can be anything from violent to blissful. Violence and discord flourish when the agreements are based on fear or self-serving mentalities, while harmony ensues when beings see the loving cores in each other. Monroe believes not only that these dynamics are played out in physical existence, but that every mind that has ever lived is part of an information field in which relationships keep unfolding in the context of different rule sets and shared simulations.

Although most people do not consciously perceive all the realities they are linked to, the theory of the collective unconscious suggests that every one of us is engaged in a process of uploading and downloading from a collective data set. This would explain why many mystics say that the solutions we find within ourselves are also solutions for all of humanity — because every human can access the patterns created by a single human mind. This hypothesis isn't yet widely accepted, but it's useful in trying to make sense of the experiences that can occur during deep meditation.

During one of my meditation trips, I found myself completely unidentified with my personality. Instead I perceived a vast yin/yang symbol full of conscious beings. Though the symbol wasn't shaped like the traditional Chinese version, it was a visual representation of the same idea. I knew that what I was seeing was the merging of opposites, and the opposites that were merging appeared to be beings from entirely different information systems, universes, dimensions, or reality simulations. I understood that I was their project and that my psychological development was their merging process. This was the most painful ego death experience I've had. My whole personality and everything I thought was real in this life vanished, and all of my consciousness was in this infinitely vast sea of merging opposites. This was very overwhelming, because the idea that my psychological patterns were part of a multidimensional evolution was beyond my ability to integrate and because of that they were disorienting.

The perceived truth within this experience was so profound that it forced me to research the idea of the collective mind. Most of the research I found in that area speaks about the collective mind or collective psyche as something that only pertains to this planet. However, during the experience I had, this planet and the forms that have evolved here were only a very small part within a much larger field of evolving consciousness which we were also contributing to. In fact, some of the beings even told me that I was one of them, that I was an alien and that I had identified with a human body for a particular purpose.

Perhaps striving for inner liberation isn't just a human process. Perhaps the psychic patterns we create have multidimensional and intergalactic significance. Maybe the stories of our lives are created by a larger pattern which attempts to merge different kinds of consciousness. Perhaps our inner struggle isn't only personal, but is a necessary process within the evolution of the collective mind. From this perspective, self-actualization is the harmonization of all the different information fields that have come together to form our personalities and our human experiences.

CHAPTER 30:

The Healing Journey

Different interpretations of a sudden and turbulent growth phase

A healing journey, sometimes referred to as an initiation crisis, is a process in which a person goes through altered states of consciousness and eventually awakens to a larger reality which dissolves fears and rearranges the person's psyche in a life-affirming way. I am using the terms *healing journey* and *initiation crisis* very broadly because the process can be very different across various personalities and cultural backgrounds. If a person has a lot of fears and strong resistance to mystical experiences, the term *initiation crisis* might be more applicable, since letting go of strongly held beliefs can be a very tumultuous process. People who experience a smoother ride might prefer to call the awakening experience a healing journey.

There are many frameworks that attempt to conceptualize the awakening experience and the emotional crisis that can come with it. One of the better-known frameworks is the Hindu concept of Kundalini. Hindus believe the Kundalini to be a divine serpent that lays dormant at the bottom of the spine and gets awakened through spiritual exercise such as meditation and yoga.

I have met many people that claim to have had the Kundalini experience. Apparently it is a very intense experience that can drive you temporarily mad if the energy isn't allowed to flow properly. My friend told me that when she had the experience it felt like a snake was winding itself up her spine and then shot out of the top of her head. But every person who told me about their Kundalini experience knew about it beforehand.

My hypothesis is that the Kundalini isn't actually a divine serpent, but is in fact the same energy that transforms the body during all other awakening experiences. It presents itself as a Kundalini experience for certain people because this framework allows a person to surrender to something known and trusted. For example, the energy might be experienced by a Christian as the touch of Christ. A friend of mine who said he was transformed through such an encounter reported seeing Christ in a vision; in that vision, Christ gave my friend a hug that sent a shockwave through his whole body.

Transformative experiences in all religions and mystical traditions tend to take on the form that is most digestible for the recipient. If an alien-looking creature tried to give a Christian a divine hug, it would likely strike the Christian as a very disturbing experience. On the other hand, someone who strongly believes that there are benevolent aliens out there would welcome such an encounter. But what if an awakening experience comes to a materialistic atheist who doesn't believe in anything beyond the physical world? What metaphors would allow for an energy transfer in that case?

A materialistic scientist who doesn't believe that there is a nonphysical reality would most likely consider all aspects of an altered state of consciousness to be pathological. Out-of-body experiences and spirit communication would fall into the category of hallucinations or symptoms of schizophrenia. This is why it is often difficult to follow the invitation to a transformative awakening in the context of Western culture. I know many people who experienced the beginning of an initiation crisis, but didn't have a supportive belief system or social environment that allowed them to trust the process.

The psyche is like a house, and when one starts to make changes to the foundation, the roof and the walls might shake. This frightens the Western mind, especially when it doesn't have access to an applicable intellectual model.

When my first awakening experience occurred, I was an atheist who believed in the doctrine of materialistic science. It was very hard to go through the process while having no belief in a nonphysical reality or a transcendent intelligence.

Transcendental experiences can be very scary for atheists, because in the atheist worldview such experiences are essentially viewed as brain malfunction. I learned that I could only go forward by trusting this transcendental intelligence as a guide and teacher. Although I couldn't put a name to this intelligence, I could tell it was reworking the energy flow of my body, and I started to believe that it came from beyond my three-dimensional reality and was much more conscious than I was.

Awakening experiences often rework your energy field to hold another level of consciousness. Your being and all of its subconscious forces are rearranged so that they become aligned with the unfolding of your highest potential. This can feel like a snake winding up your spine, or like laser beams shooting into various points in your body, but it is ultimately a deeply individual process that speaks to you in a metaphorical language. This language can involve words, symbols, images, feelings, or sensations. It all depends on how you have chosen to interact with the human psyche and how your behavioral patterns can be aligned most efficiently.

The core element of a healing journey is an altered state of consciousness through which the psyche optimizes itself. This can happen through many different types of experiences and in many different contexts. For example, in a 2013 TEDx Talk, Anita Moorjani told her story about healing herself from cancer with the insights she gained during a near-death experience. She was dying from end-stage lymphoma, she had tumors the size of lemons throughout her body, and her doctors told her family to say their goodbyes.

Anita fell into a coma that she wasn't expected to awaken from. But while she was lying in her hospital bed, supposedly awaiting her death, she left her body and had a classic near-death experience. She said that during the experience she had 360-degree vision, and she could not only see everything around her body but could also feel what everyone was feeling in the room. Her awareness wasn't limited by time or space, either; she said she simultaneously perceived her brother, who was in an airplane coming to see her.

Anita experienced many of the classic signifiers of a near-death experience: going into a bright light, meeting deceased relatives and so on. She said that right before she entered her body again, the spirit of her father told her: "Now that you know the truth of who you really are, go back and live your life fearlessly." Anita had the distinct realization that her awareness was one with the awareness that everyone else shared; with this realization, she said, she no longer feared death.

When she opened her eyes, the doctors were surprised. In the following days, they were shocked as her tumors began to disappear. Within a few weeks, she left the hospital with no sign of cancer.

The doctors couldn't explain Anita's recovery, but to her, it was crystal clear that coming to an understanding of who she really was had transformed her illness. She experienced herself as eternal awareness, connected to all that is, and this experience was so profoundly transformative that it reversed her terminal cancer.

After she was released from the hospital, she was frequently invited to medical conferences where she talked to cancer researchers about her experience, but she stopped going when she found that the scientists there didn't take her claims seriously.

Emotions are the language of our cells. The difference between two emotions isn't only a subjective experience. Our body produces different hormones and enzymes depending on how we feel, and our cells react to these hormones and interact differently with our genes. Our emotions determine which genes our cells choose to express. These findings have been documented by scientists like Bruce Lipton and Dr. Joe Dispenza. Who knows how far this cellular communication reaches? Who knows how intelligent our cells really are?

This wasn't part of my research for this book, but while on tour I got the chance to talk to a friend's aunt who happened to be a cancer researcher. I asked her if cancer occurs when cells act selfishly. "No," she said, "cancer cells aren't selfish. They share nutrients with each other. Cancer occurs when cells forget that they are part of a larger whole."

Perhaps Anita was healed because her mystical experience provided her with the experiential knowledge of a larger reality. It is one thing to think about the interconnectedness of consciousness, and another to experience it directly. When you experience it directly, you believe it, and it affects your emotions and all cells in your body.

The difference between an initiation crisis and a philosophical consultation is the difference between experiential knowledge and rational knowledge. Rational knowledge is divisive and everything needs to be broken down into relationships between subject and object. Emotional, experiential knowledge can be uniting; you can feel at one with a tree, for example. Such an experience doesn't make any rational sense, since you and the tree fall into two different categories. However, in the world of feeling, the boundaries between subject and object no longer need to exist. Oneness can be experienced, and that experience can profoundly affect a being.

In a free will system, the concept of "health" is the collaboration between all its agents. When we recognize the larger reality and integrate it into our being, we join a current of consciousness that extends beyond the intellectual idea of our isolated self. We join the collaborative whole when we connect to a higher power with love; this is the whole purpose of a healing journey or initiation crisis. But the process of connecting to a larger whole doesn't need to involve a crisis.

Some people deepen their connection through very pleasant mystical experiences. These experiences only become a crisis when the person experiencing it is invested in the belief system of his previously isolated personality. In this situation, a mystical experience triggers an unraveling that can be scary, as the illusions that our separated self created need to be confronted.

My first mystical experience was followed by extreme doubt and fear. I wasn't able to surrender to the great mystery. Instead, I tried to use my intellect to make sense of everything I was experiencing, which felt like driving on the highway in first gear. Emotions can process a lot more data than the intellect, which is why spiritual teachers encourage their students to feel emotions fully without being distracted by thoughts. By fully submerging your attention in the experience of your emotions, you can process information more quickly.

A rational-minded person might need an initiation crisis to realize this, but it can be extremely difficult to follow a process that isn't widely accepted in Western culture. People going through an initiation crisis often end up in mental hospitals because no one in their social environment can guide them through the process.

I interviewed Phil Borges, director of the 2017 documentary Crazy-Wise, about the attitud*es indigen*ous tribes have toward events that we in the West consider to be mental breakdowns. I was sitting with him on his porch in Seattle when he told me how he ended up exploring the relationship between spiritual awakenings and mental illness. Borges told me that he started exploring the topic when he was working as a photographer for the United Nations. He traveled around the world and documented shamans from a wide variety of tribes. Borges noticed that the behavior of a shaman going through an initiation often resembles what we in the West consider to be a psychotic episode. The only difference is that other cultures consider the symptoms to be good news. The village celebrates the arrival of a new healer, and the young healer is taught how to integrate his gifts by a tribe member who already went through the process.

This process varies from tribe to tribe and from person to person. Sometimes it takes weeks, months, or years for the person to grow out of his psychosis and become a healer.

In the West, psychotic breaks are considered to be pathological conditions that need to be treated with medications. But if a psychotic break is actually an initiation crisis, one needs to learn how to deal with the larger reality and work with the symbols, voices, or nonphysical beings involved to unite all aspects of oneself. This is hard to do when friends and family are worried and doctors tell you that what you experience are symptoms that need to be treated.

In contrast, the shamanistic worldview sees the symptoms of the patient as the unfolding of the healer-to-be's purpose. When a person sees immaterial beings, they are encouraged to form relationships with them, and if these relationships aren't harmonious, they either have to be ended or transformed. The person's experiences are a process of forming harmonious relationships with the conscious information fields or beings they interact with. I believe that many people in Western society still go through the initiation crisis without knowing about it and without having a support system that sees their symptoms as signs of an awakening.

In Stephanie Marohn's article "What a Shaman Sees in A Mental Hospital," Marohn recounts the experiences of Malidoma Patrice Somé, an African shaman from the Dagara tribe who visited his friend in an American mental hospital.

"I was so shocked," Somé said. "That was the first time I was brought face to face with what is done here to people exhibiting the same symptoms I've seen in my village."

According to Somé, the Western doctors he observed took the exact opposite approach than the elders of his villages. His tribe saw mental illness as the process of forming a relationship with beings from another world to bring a message to humanity and help humans expand their consciousness. But instead of helping the person go through the necessary transformation, the Western doctors saw the symptoms as problems that needed to be stopped with medications.

"This is how the healers who are attempting to be born are treated in this culture," Somé told Marohn. "What a loss! What a loss that a person who is finally being aligned with a power from the other world is just being wasted ... The Western culture has consistently ignored the birth of the healer ... Consequently, there will be a tendency from the other world to keep trying as many people as possible in an attempt to get somebody's attention. They have to try harder."

Shamans view insanity as a transitional phase. To allow a person to reach a new level of consciousness, the symptoms need to be viewed as guidance and not as signs of mental illness. When we view them as signs of mental illness, we don't allow ourselves to follow the healing process that can occur during an altered state of consciousness.

I also talked to Tom Campbell about this topic, and he told me that mental instability can be a temporary phase in which an older soul goes through a rapid process of growth to catch up with where its evolution left off in a previous life. Campbell actually told me this in regards to my own experiences. I first reached out to him because I was trying to make sense of some experiences I had in nonphysical realities. Before I started this book, I had never heard of anyone having these types of experiences, which made them difficult to integrate. But what initially was a terrifying and disorienting experience later became a source of trust, inspiration, creativity, and love. It is a shame that many people go through similar transitions and are locked up in mental hospitals where their symptoms aren't recognized as the awakenings that they might be.

CHAPTER 31:

Unrecognized Awakenings

A shamanic view of mental illness

When spiritual awakenings aren't correctly integrated, they can become spiritual emergencies, which might resemble mental illness but require a very different kind of treatment. I don't know enough about mental health to say where exactly to draw the line. Are all mental illnesses spiritual emergencies that are misunderstood, or are there illnesses that cannot be healed through spiritual integration? These are important questions to explore, but are beyond the scope of this book. Still, it's clear that taking a different perspective on the nature of reality necessarily requires taking a different perspective on what "mental health" means.

If reality is a product of material interactions, then mental illness must be the result of a chemical problem in the brain. If we look at it from the perspective of a consciousness-centric world view, though, then the brain is an information structure that can be reworked within the realm of consciousness. The latter is also supported by the findings of neuro-plasticity, a research field in neuroscience that explores the brain's ability to change based on what we feel, think, and do.

Whatever a person encounters in the realm of consciousness also affects the brain, and any method that pulls seemingly separate psychic forces together has the potential to heal. In meditation, this is done through the common aspect of awareness. All phenomena in the psyche have one thing in common: you are aware of them. They happen within the context of an aware being. Noticing this can bring a fractured psyche back together and induce healing or coherence. I think this is why scientists like Richard Davidson and Joe Dispenza have found that meditation has physical and measurable effects on the brain. Although meditation is essentially an exercise within the realm of consciousness, the healing that occurs there has physical consequences.

Healing mental illness doesn't mean reverting back to a normal state of consciousness. Hearing voices or seeing visions can be a terrifying experience, but it can also be an invitation to step into one's full potential as a person and a healer. When you heal yourself by acknowledging a larger reality and finding peace in your awareness, you can share the liberation that arises through anything you do.

My friend Joro is a DJ in New York. We used to host events together that aimed to promote alternative ways of looking at mental health and mystical experiences. When he was 40 years old, he encountered nonphysical beings for the first time. At first he thought he was going crazy, but somehow he managed to go through the process with trust and, like me, found that these experiences became positively transformative. I decided to include his story because it highlights the highly individualistic process through which these experiences can interact with our physical systems.

This is how Joro described his experiences to me:

> *When the experiences first started, there were three entities. One had a very sexual energy. It delivered blasts of transformation that I experienced as anal orgasms. The*

intensity of the physical experience itself was transformative on its own - I had never experienced such an overwhelming degree of sensation. It was beyond pleasure, a direct opening of the physical dimension of existence to the energy that is behind everything. In parallel, the entity would deliver blasts of transformation and knowledge on all scales from a very personal (who I am, how I am, what am I doing here), to the level of society, species, planet, up to the level of the universe and existence in general. I did not have a name for this entity.

Another entity felt quite intense and unpleasant (I used to call him "the Evil One"). It felt not human, but animalistic like a wolf, very aggressive and dominating, and when he would appear, it would feel like a possession. Still, I remained open and receptive to the experience, allowing it in its fullness. I noticed that if I didn't resist it, it would "establish a dominion" and that would spend its energy and purpose. Establish a dominion felt like something coming into me with the mission of conquering. The Wolf One came with such an overwhelming force that there was no chance of resisting (even though that was my spontaneous first reaction). Instead, I trusted the experience and allowed it with a curious and open mind. That made all the difference - it was a lesson of letting go and not judging first but being open and learning. Because of that openness, I cultivated a relationship over time so that I could participate in setting the boundaries of my education. Learning in general is quite often what we would consider "a negative experience": burn your hand, learn to avoid fire; fall from your bike, learn to ride. The "negative voices" have their own function and purpose.

So the third entity would come in, with a different purpose. I used to call her "the Motherly One". She would come with an intention of care, and the understanding I would get through her was usually the deep mystical experience of connection. She would deliver the highest knowledge and in a way pull together everything that the three entities communicated.

They wouldn't necessarily appear in this order. Over time, I stopped calling the second one "the Evil One" and referred to him as "the Wolf One" (on advice from an Ayurvedic doctor from Guatemala.) As they would say, good and evil are human perceptions and they are irrelevant for this type of understanding and experience. Gradually more entities started appearing and the original three faded (only the Motherly One remained fairly regular). Sometimes they would set up tests about my commitment to the experience, testing my limits (e.g. "What if we told you to kill yourself?" Or "What if you died right now?") I learned how to communicate with them and to discern when to follow, when to draw a line, etc. This has been profoundly meaningful and had made an impact on everything from my behavior to how I communicate with people.

They have also delivered warnings numerous times. First, this experience should not be taken lightly, it is not entertainment. Second, even though my task is to share this experience, I should keep in mind its subversive potential that many will find dangerous. To deliver this type of message in an effective way, one shouldn't just indiscriminately blast it out but learn an appropriate practice of communication. This type of practice of life and communication (living in truth) over time aligns one fundamentally with everything and that becomes powerful, not ego power but the power of life and love.

Eventually the communications lost what would be considered "negative voices" and became more straightforward, so I began recording them.

Joro used to attend meetings with other people who heard voices to share his experiences. But the idea that one could develop a positive relationship with the voices and learn from them wasn't part of their accepted framework. It is rare to find mainstream psychologists that see the experience of nonphysical beings as beneficial. Not many dare to oppose the status quo and explore their own psyche on that level of depth, but there are exceptions.

Carl Jung didn't just study psychology theoretically — he also dedicated his life to exploring his own psyche. Visions became his avenue of psychic exploration. In early 20th-century writings later published as *The Red Book,* Jung details dialogues he had with immaterial beings. At points he doubted his own sanity because the depths of his exploration surpassed the psychology of his time. Ultimately, though, Jung's self-exploration led him to a new understanding of mental health.

From *The Red Book*:

When the desert begins to bloom it brings forth strange plants. You will consider yourself mad, and in a certain sense you are mad. To the extent that the Christianity of this time lacks madness, it lacks divine life. Take note of what

> *the ancients taught us in images: madness is divine. But because the ancients lived this image concretely in events, it became a deception for us, since we become masters of the reality of the world. It is unquestionable: if you enter into the world of the soul, you are like a madman, and a doctor would consider you to be sick…*
>
> *…If you do not know what divine madness is, suspend your judgment and wait for the fruits. But know that there is a divine madness which is nothing other than the overpowering of the spirit of this time through the spirit of the depths. Speak then of delusion when the spirit of the depths can no longer stay down and forces a man to speak in tongues instead of human speech, and makes him believe that he himself is the spirit of the depths. But also speak of sick delusion when the spirit of the time does not leave a man and forces him to see only the surface, to deny the spirit of the depths and to take himself for the spirit of the times. The spirit of the times is ungodly and the spirit of the depths is ungodly, balance is godly… Whoever does the one and does without the other you may call sick since he is out of balance.*

In these paragraphs, Jung defines sanity as a balance between the forces that govern surface reality and those forces that surpass reason and logic. His inner work was focused on creating balance, but not through the force of his will. Jung's willingness to surrender to his soul and allow his soul to take him on journeys is a constant theme that runs throughout *The Red Book*. These journeys were often frightening and painful, but his willingness to follow his soul all the way allowed him to complete the process and bring back the fruits as the wisdom he shared with the world.

Again from *The Red Book*:

> *In the following night the air was filled with many voices. A loud voice called, 'I am falling.' Others cried out in confusion: 'Where to? What do you want?' Should I entrust myself to this confusion? I shuddered. It is a dreadful deep. Do you want me to leave myself to chance, to the madness of my own darkness? Whither? Whither? You fall, and I want to fall with you, whoever you are.*
>
> *I see a gray rock face along which I sink into great depths. I stand in black dirt up to my ankles in a dark cave. Shadows sweep over me. I am seized by fear, but I know I must go in. I crawl through a narrow crack in the rock and reach an inner cave whose bottom is covered with black water. But beyond this I catch a glimpse of a luminous red stone which I must reach. I wade through the muddy water. The cave is full of the frightful noise of shrieking voices. I take the stone, it covers a dark opening in the rock. I hold the stone in my hand, peering around inquiringly. I do not want to listen to the voices, they keep me away. But I want to know. Here something wants to be uttered. I place my ear to the opening. I hear a flow of underground waters. I see the bloody head of a man on the dark stream. Someone wounded, someone slain floats here. I take in this image for a long time, shuddering. I see a large black scarab floating past on the dark stream. In the deepest reach of the stream shines a red sun, radiating through the dark water. There I see - and terror seizes me - small serpents on the dark rock*

> *walls, striving towards the depths, where the sun shines. A thousand serpents crowd around, veiling the sun. Deep night falls. A red stream of blood, thick red blood springs up, surging for a long time, then ebbing. I am seized by fear, what did I see?*
>
> *...Depths and surface should mix so that new life can develop. Yet the new life does not develop outside of us, but within us.*

The essence of this experience isn't in the metaphors and symbols, it's in the emotions Jung felt and the insights that he gained; in other words, the psychic energy that was released through the experience. This psychic energy contributed to an integrative process. The passage "depths and surface should mix so that new life can develop" reflects a common theme among shamanic journeys (or whatever you choose to call these internal trips).

When my experiences started, I didn't know much about mythology and visualization exercises, and I didn't experience symbols like caves, beetles, and beheaded heroes. But the psychic energy that began to flow through my body resembled the energy that these metaphors represent.

My process also put me into states that resembled insanity. I remember having conversations with several nonphysical beings. Some of them seemed like the spirits of the depths, very dark energies, and others seemed to be very enlightened; it was as if they were two different types of civilizations.

I could sense that these two opposing energies wanted to merge or create some sort of bridge between them. The enlightened beings — spirits of love and light, as I experienced them — told me that they wished to use my body to create a new information system through which the flow of consciousness that was stuck in the identity of evil spirits, or more fear-based beings, could be liberated. They showed me the image of a blue figure eight, which they explained was the end result of this project. The first time these beings approached me I couldn't complete this project, because the psychic forces were stronger than my ability to balance their input. But a year later they approached me again with the same invitation.

I was very curious about the nature of these beings, and whenever I asked questions I seemed to receive answers. It appeared to me that one of the civilizations I connected to was using the energy of suffering to fulfill its own needs for love. Somehow this civilization was funneling the energy released when another being suffers and building vast networks with that energy. Not unlike humans do with factory-farmed animals. In this case, though, humans were the animals and our suffering was the crop of this inter-dimensional farm. The beings that had evolved within this parasitic civilization were actually very loving, though, just like humans who eat meat while completely unaware of the suffering their consumption is causing.

The beings of love and light were connected to the source directly, and that meant that they truly embodied pure love. They wanted to give instead of take.

The project these beings wanted to work on was the restructuring of a larger network so that the parasitic civilization's energy flow united with pure love. This meant that their survival would no longer be dependent on creating suffering in others. I couldn't understand it much further than that, and I had no way of knowing whether the metaphors my mind was creating were accurate.

When I asked how I was going to be of assistance in this project, they told me that I needed to lend them my physical body, including my focus. I agreed. For the following weeks it felt like laser beams were creating geometric structures in my brain and body, and that tremendous amounts of energy were flowing through these structures. This process went on and on as I tried to keep my mind calm and focused.

Eventually this energy began to accelerate into a rapid flow. I suddenly found myself with no bodily awareness, accelerating in the shape of a figure eight. Within this figure eight I perceived the essence of my friend Mike. We both attracted and repelled each other, which increased the speed at which we traveled through this figure eight. At some point the speed appeared to be infinite and all that remained was a glowing figure eight.

I gradually came back to my bodily awareness, through which I could feel the cheering of millions of souls that all had participated in or anticipated this process. It felt as if the depths had been merged with the surface and a new platform for life had been created, in which countless beings were united for the purpose of liberation.

This process had gone on for several weeks. Before it was completed, there was a moment in which I could see with my closed eyes that a light beam had come out of my heart. As I followed this beam, my vision led me to another reality in which I met Carl Jung, who congratulated me on my progress.

I believe that the purpose of these kinds of experiences isn't to lift the stone, witness the serpents, or talk to this being or that being. The purpose is to build the network within consciousness that is created through the experience. When you encounter parts of yourself, or parts of the collective mind that frighten you, you also encounter a membrane that prevents information from flowing. An encounter with that which scares you gives you the opportunity to rework it, to trust the fall, to climb into the cave, to complete the project. These are the choices that liberate fear and allow it to dissolve into its original energy, which is love consciousness — the divine in action, as some mystics refer to it.

I think that shamanic journeys are experienced as real for the same reason that our life is experienced as real. We can only move psychic energy through the power of belief. When something is experienced as real, we believe it, and our belief moves the energy that the metaphors represent.

A lot of the deep work I have done in the collective mind seemed to be related to the split caused by traditional Christianity and other religions that do not see the divine in all that is. When you encounter a frightening being and you believe that it belongs to Hell, you add to the split that has caused this being to grow away from the light.

I believe that the truth is that anything that has awareness is of God, and that God isn't a judgmental man in the sky, but an energy that tries to replicate love within complex systems, such as our universe, and everything within it. God isn't perfect because God is alive. Anything that is alive grows, and how can something that grows be perfect or complete?

When you see suffering, it means that this divine energy has entangled itself in layers of identity until it wasn't able to know itself anymore. So when we have experiences involving what our culture imagines as Hell, it is important to remember that anything we see in or from this place is actually divine, but has forgotten its own nature. Holding on to this knowledge transforms fear into compassion.

While doing research for this book, I studied cases of famous historical figures descending into insanity. What I found is that many brilliant minds, including Carl Jung, Friedrich Nietzsche, and Issac Newton, were considered to be crazy at some point in their lives. Carl

Jung managed to get through his madness by integrating the psychic energies he encountered. Unfortunately, Nietzsche and Newton did not find a way to ground their experiences, and they died as madmen. (There are some theories that Nietzsche lost his mind because of syphilis and Newton due to mercury poisoning, though there's no consensus among historians who studied their lives.)

After I read up on the nature of their experiences, I began to think that these men might have gone through an awakening which they never integrated. An awakening is driven by a psychic force that reworks your personality and beliefs through whatever process our individual nature requires. The personality and the intellect get restructured and become aligned with a larger vision that includes the divine or transpersonal. However, these experiences are frightening and disorienting when the work of an entire life gets re-contextualized, and all that has been thought and believed is suddenly understood from a more expansive vantage point. Old beliefs break in the face of incomprehensible experiences, which leaves the person temporarily disoriented. I believe that some of Nietzsche's "madness letters" are clear signs that he was going through an awakening.

In a letter to composer Richard Wagner's widow Cosima Wagner, Nietzche wrote the following:

To Princess Ariadne, My Beloved.

> *It is a mere prejudice that I am a human being. Yet I have often enough dwelled among human beings and I know the things human beings experience, from the lowest to the highest. Among the Hindus I was Buddha, in Greece Dionysus — Alexander and Caesar were incarnations of me, as well as the poet of Shakespeare, Lord Bacon. Most recently I was Voltaire and Napoleon, perhaps also Richard*

> *Wagner ... However, I now come as Dionysus victorious, who will prepare a great festival on Earth ... Not as though I had much time ... The heavens rejoice to see me here ... I also hung on the cross ...*

At first glance, this letter sounds like sheer madness. If we take a closer look, though, we can see that Nietzche was dealing with the transpersonal aspects of awareness.

During altered states of consciousness, our isolated human identity is experienced as illusionary. Integration occurs when both the small/isolated self and the transpersonal self can coexist in harmony. Nietzsche seemed to have lost his connection to his human identity and recognized himself as the current of awareness that flowed through several enlightened beings. This is a state of consciousness sometimes experienced during deep meditation.

Some channeled texts suggest that many beings exist as a collective whole after their physical lives. Paul Selig's *The Book of Mastery*, a text which Selig claims to have channeled from non-corporeal spirit guides, talks about Christ consciousness as a state of consciousness that consists of many beings. According to Selig, the beings that speak to him say that personality and separate identity are like clothing which they can choose to put on or take off. When we have physical bodies, though, we need to respect the illusion of separation to continue to function in a three-dimensional reality.

When an awakening isn't integrated, it is experienced as madness because the collective current of shared identity doesn't have a fitting vessel. A fitting vessel is a body and a psyche that has the emotional and intellectual capacity to hold this energy. But when the body and psyche have been excessively conditioned by a culture that has no concept of collective awareness, many dams need to be broken, and that is often not a smooth process. The true self or soul can suddenly no longer identify with what it has created in the context of one human life.

Toward the end of his life, Issac Newton seemed to be going through such a process. Although he was mostly known as a mathematician and the founder of the materialistic world view, he actually spent more time writing about alchemy and religion than about mathematics and physics. Newton believed God was an all-pervasive being that extended beyond human imagination and that the physical laws were part of God's will: "The most beautiful system of the Sun, planets, and comets," Newton wrote, "could only proceed from the counsel and dominion of an intelligent and powerful being."[26]

Like Nietzche, Newton wrote letters in his late years that suggested he had gone mad. According to his correspondence, he believed that he was a divine messenger who was supposed to bring a gift to humanity. He certainly did, by giving us the concept of gravity and the mathematics to calculate the planetary relationships within our solar system, but he also wanted to add a spiritual component to the newly born worldview of materialism. In this he never succeeded.

Sarah Dry's 2014 book *The Newton Papers* tells the story of the scientist's mystical writings, which were suppressed, hidden and forgotten for almost 300 years until finally seeing publication in 2007. By then, though, humanity had advanced far beyond the frameworks of medieval alchemy, and few scholars were particularly interested in excavating what Newton had to say about the subject.

I wouldn't be surprised if Newton's mental breakdown was caused by an integration issue. Even while he was laying the groundwork of materialism, he was a deeply spiritual man. He was giving birth to a

[26] PBS, "Newton's Dark Secrets."

new paradigm, but wasn't able to add the other half. Afraid of backlash from the Church, he kept his work on mysticism a tightly held secret.

It is often difficult for people who work on the depths of their own psyche to handle the fact that society would consider them insane if they openly spoke about the work they are actually doing. The stress of living in such a way can easily develop into fear and paranoia.

Toward the end of his life, Newton was very paranoid because he knew that the information he possessed was challenging the Church's teachings on multiple levels. I could imagine that this fear added to his inability to make peace with the psychic forces that he had conjured. It is much more difficult to integrate an awakening if you don't feel safe in your society or social surroundings.

We still face this issue today. When a person experiences a layer of reality that our society isn't comfortable with, it tends to frighten and concern their families and doctors. We don't have to fear the Church trying to burn us at the stake, but the fear of being locked up in a mental institution or of being ostracized by one's family still can be an immense obstacle to integration. It is hard to let your small self embrace your transpersonal self while you feel threatened, because the fight or flight instinct pushes you into stress and away from coherence, balance, and peace.

It might take a few more decades until mainstream medicine is ready to embrace the mystical. But imagine what it would be like if a soul that has incarnated to challenge the status quo was not ostracized but embraced. Imagine if we had institutions that would allow a soul to go through the process of reworking its psyche in a safe, open-minded facility. I believe we could reduce mental illness dramatically if we recognized it as a spiritual emergency that demands integration, not medication.

CHAPTER 32:

Blessing Fear

Learning to see the origin in that which frightens us

Fear is the consciousness modality that prevents growth and stifles the awakening process, because fear is focused on preservation and not expansion.

Fear can express itself as any self-serving emotion, such as greed, anger, or jealousy. These psychological variations are part of the same spectrum of self-serving potentials.

When we encounter other people who embody these emotions, we tend to write them off as evil or bad humans. We think that we need to protect ourselves from such people, we think that such people deserve to be punished, and we think that the people we care about should be afraid of them. We feel safe by acting on our fears, but our reactions are just perpetuating the cycle.

The cycles that fear creates can only be broken when you give fear-based actions love-based reactions, when you bless someone who has hurt you, when you feel compassion for those who have wronged you, and when you have patience with the greedy and selfish. It is easier to do this when you realize that anyone who follows self-serving emotions has simply forgotten who they are.

When you encounter fear-based emotions during your meditations or your out-of-body journeys, the same principles hold true. If you are trying to protect yourself with force, you are responding to fear, which empowers that energy. You most effectively liberate fear when you remain anchored in your understanding of your true self or love. Fear makes you identify with thoughts, but your true nature sees all identifications as passing phenomena. Consequently, you let thoughts flow when you know your essence, and you cling to them when you forget who you are. In this wider space of identity, you can have terrifying experiences without being terrified. Fear is reduced to an energy and its content becomes irrelevant.

During one of my meditations, I saw the face of a witch flashing in front of me. Flesh was hanging from her skull and her eyes seemed to be radiating pure evil. At first I was startled, but then I understood the deep misunderstanding that was driving this vision. Whatever information stream I was connected to was representing a relation within consciousness that had been organized by fear. Whatever being or psychic symbol I had encountered had suffered to the point where it had given up all hope of ever seeing the light again. This understanding filled my eyes with tears, and I felt a tingling energy enter my heart. It felt like my heart was a radiant light that welcomed a part of itself that it hadn't seen for a long time. This was my subjective experience, but it was enriching because it allowed me to encounter a terrifying symbol with compassion and love.

According to Rupert Sheldrake's theories we previously discussed, all events, beliefs, and cultural habits create information fields. If you look at Christianity from this perspective, it would make sense that the fear of the devil, the fear of witches, and the fear of demons gave these ideas power.

Christianity was spread violently and any culture that worshipped different gods or spirits was thought to be worshipping the Devil. This belief gave Christians the justification to murder and enslave tribes across the globe. To say the least, these actions were non-integrative, creating a divide between the Christian information field and that of the cultures they murdered and enslaved. The tribal gods and spirits can be seen as conscious networks that have been growing within these tribes for hundreds of generations. Suddenly the symbiosis of physical and nonphysical beings was interrupted by invaders that believed anything that differed from their dogma comes from the Devil. This created a divide in the collective consciousness that still needs to heal.

If we have demonic experiences, we may be connecting to an information field that has suffered in the past. We can only contribute to the liberation of these fields when we don't react with fear, but rather focus on seeing the origin of this fear-based pattern. The fear of the unknown and the inability to tolerate diversity has created much suffering and anger. Recognizing past pain and encountering it with love is what will help heal the collective trauma of humanity.

CHAPTER 33:

Drug-induced Awakenings

Being launched with a cannon instead of taking the stairs

Every summer my band plays at large festivals in Europe and North America. Sometimes we hang out at the festivals after we have played. I find it interesting to interact with people that immerse themselves in music and drugs.

One time I was somewhere in Ohio at a large festival and I had a conversation with a young woman. When I asked her what she had taken during the weekend, she rattled off a remarkable litany: LSD, mushrooms, MDMA, cocaine, ketamine, marijuana, and alcohol. When I asked her why she takes drugs, she said: "Because they make me feel free. When I'm sober, I'm a very uptight person."

It seems that many people take drugs for similar reasons. People want to escape the limitations of their personality and do it in different ways and to different degrees. Some are content when they feel a bit more loose, while others want to feel as though they've exited physical reality entirely.

I have talked to people that told me that they often meet nonphysical beings or guides when they take hallucinogens. The states of consciousness people experience on hallucinogens seem to resemble the altered states of consciousness that meditation and other mind-body practices can induce. For many people, taking psychedelics is a way to experience the state of oneness that has been known to Eastern traditions for thousands of years. Meditation deactivates the brain's default mode network, which is a neural circuit that is also deactivated by psychedelics.

The deactivation of the default network is a model that many researchers now use to make sense of mystical experiences and altered states of consciousness. The default network is a neural circuit that provides us with the content of our worries, self-doubts, and other worldly concerns. It has evolved to help humans determine our relationships with the external world. When it is deactivated, either chemically or by exercising focus, we can have transpersonal experiences that aren't limited by our normal sense of self.

When our consciousness is no longer busy receiving the data of the default mode network, it is free to experience the deeper aspects of awareness. But this shift isn't always smooth — especially if one doesn't know that there is such a thing as awareness beyond personal identification.

Many people live their whole life within their default mode network and believe that their worries and self-images are reality. When they experience the effects of psychoactive chemicals, everything they thought was important might be suddenly reduced to a cosmic joke. This can be very liberating and freeing if one is willing to let go, but not everyone is prepared for this.

At one festival in Florida, I walked around and found a tent hosting audience members who had overdosed or were stuck somewhere in their minds, struggling with the mystical. But instead of providing a relaxing environment and encouraging them to let go, the area was exposing these people to the flashing lights of ambulances and the high-stress environment of a medical emergency facility. It is important to have medical help in the proximity of a festival, because some drugs are physically dangerous. However, with psychedelics, the problems are almost always psychological or spiritual. When people aren't able to let go of their small selves while experiencing the energy of their souls blasting through their heads and hearts, it creates a dissonance that can resemble insanity. These people are actually experiencing spiritual emergencies that need to be treated with calmness, love, and wisdom. Psychedelics can have a devastating effect on people — not because they're inherently harmful, but because our culture has no framework for mystical experiences.

I've experimented with psychedelics myself and found that they give you a taste of what intense meditation can do. The difference is that you don't go into the altered state gradually, but are shot into it with a cannon. Still, it's essentially the same place — a beautiful place of unity that can be terrifying if you resist it.

Michael Pollan addresses this in his 2018 book *How to Change Your Mind.* He explores how our culture has misunderstood these drugs and has failed to create a space for their beneficial usage. Within the last decade, though, things have been changing as the medical industry slowly begins to recognize the potential benefits of psychedelics. There have been hundreds of studies on LSD and psilocybin that explore their health benefits, and the College of American Pathologists' website at cap.org hosts a huge collection of psychedelic research.

When psychedelics are taken in a safe and comfortable space with the assistance of a therapist who encourages the patient to let go, they can cure or improve a huge range of mental conditions ranging from depression to anxiety, addiction, and post-traumatic stress disorders. A February 2019 article from Wired, "Inside The Push To Legalize Magic

Mushrooms For Depression And PTSD," features the story of Todd, a cancer patient who struggled with these issues during his treatment.

"My mental condition was deteriorating rapidly," Todd told Wired, "and I was on [antidepressant] medication No. 14 and it wasn't working… My psychiatrist said, 'I honestly think you're a big candidate for psychedelics.'"

At first Todd thought that his psychiatrist was joking, but when he followed his advice and began to trip weekly on a large dose of mushrooms, his depression and PTSD evaporated. One strong mushroom trip a week seemed to be the cure he was looking for.

Psychedelics are increasingly recognized by the medical establishment as a cure to a wide range of mental illnesses. The psychedelic research unit at Johns Hopkins Medical Center in Baltimore, Maryland has conducted a rigorous scientific study that demonstrated that psilocybin, the active component in magic mushrooms, can induce mystical experiences that have long-lasting health benefits.

Surprisingly, project researcher Dr. Roland Griffiths, a professor in Hopkins' Department of Neuroscience and Psychiatry, didn't shy away from actually using the word "mystical" in discussing the topic: "Under very defined conditions, with careful preparation, you can safely and fairly reliably occasion what's called a primary mystical experience that may lead to positive changes in a person. It's an early step in what we hope will be a large body of scientific work that will ultimately help people."[27]

I believe that the medical benefit of psychedelics is due to the larger perspective one gains through a mystical experience. As a Buddhist would say, the root cause of all mental and emotional suffering is

---

[27] https://www.hopkinsmedicine.org/Press_releases/2006/07_11_06.html

clinging to the small identity and its desires and fears. The problems of the small self are rendered insignificant in the face of a mystical experience, because in that larger context we recognize ourselves as eternal and formless awareness. During a mystical experience, this is experienced directly and isn't just intellectual knowledge. Mystical experiences give us experiential knowledge that transforms the deeper, non-rational aspects of ourselves. I believe the healing properties of psychedelics come from the fact that they can deactivate the default mode network and artificially free awareness from its identification with our personality.

CHAPTER 34:

Awakening Through Music

The non-rational language of music and its ability to expand consciousness

Music is a non-rational language through which humans communicate emotions. The structure of music is similar to a verbal language in the sense that patterns of information are passed down from one generation to another. In the case of music, these patterns are the relationships between sound waves, which are called intervals. While learning an instrument, a music student usually studies the songs of their favorite musicians; during this process, they acquire a non-rational language through which they learn to express their emotions.

A good musician has internalized a musical language to the point where his or her emotions speak directly through it. I am not sure if this happens only through sound waves, or if a nonphysical data exchange accompanies a musical experience. The energy people feel at shows could very well be a collective information field, and the sound waves could be like links that lead to websites. However the information exchange of a musical performance happens, the results are obvious to anybody who loves music. Most people use music to adjust their state of consciousness.

I believe that we like a given piece of music if the emotions that are communicated by the musician benefit our soul's growth — if the musician's non-rational language creates a state of consciousness that lifts us, fills us with encouragement, and reminds us of the present moment. This is the case because a good musician can go into a trance and be fully present with the sound he or she creates.

Learning a musical language is a type of initiation in the sense that one has to confront those aspects of the mind that prevent one from being present. To play your instrument well, you have to be able to listen to your fellow musicians while being fully present with every note you play. There simply isn't any room for thoughts, because every thought uses energy and focus, taking away from your ability to play accurately and authentically. If you are thinking about what others are thinking about you, or if you are wondering if you are playing badly, you cannot be fully present with what you are playing. So musicians are forced to learn how to deal with encroaching thoughts and how to best maximize focus. But this isn't always a conscious process. Many musicians have used drugs to alter their consciousness so that they can temporarily enter a trance.

One of my first music teachers told me there are two paths to becoming a great musician: "Either you take drugs or you meditate." Taking drugs might be easier, but doing so creates a discrepancy between low- and high-energy states. Meditation takes more effort and works more slowly, but gradually one learns to live in the flow from which one's music comes. There is no longer a difference between being on stage and walking in the woods.

These paths aren't completely separate. There are also many musicians who walk both paths — either by taking drugs moderately and mostly getting their high through meditation and yoga, or by mediating moderately and mostly getting their high through drugs. For some musicians, playing their instrument is their meditation and they wouldn't define their process rationally.

Whatever path a musician chooses, they walk a path that shamans and healers have walked in the past. The musician is a person who speaks to the community about a different state of consciousness and addresses social and political issues either through words or attitudes in consciousness. This doesn't make much sense to the rational mind, yet the energies that musicians bring to this world have continuously transformed society.

CHAPTER 35:

The Deeper Self

Dealing with mystical experiences

Before the rise of materialism, there were thousands of years of human history during which the spirit world was considered a central part of reality. But the experience of another dimension of reality hasn't gone away; the only thing that has changed is the way people treat their experiences. Although not everyone has metaphysical experiences, it isn't hard to find people who have.

Since I began to share my theories and experiences, many people have come to me and said that they've also communicated with spirits, aliens, nonphysical beings, guides, angels and the like, but never shared their experiences with anyone for fear of being thought of as crazy. The terminology is always different, but the common element is the experience of touching another reality which is inhabited by a transcendent nonphysical intelligence.

Most of the time, these experiences make an individual aware of the importance of self-actualization. These experiences can serve as catalysts that help individuals grow out of cultural limitations and claim their true identities. Every person makes sense of these realizations in different ways, but these experiential journeys are all developments toward transcendental love, which is also known as enlightenment, freedom, true self, or Zen. At the core, all of these terms emphasize a shift in identity. A person moves away from the separate ego self and begins to experience the shared aspect of awareness.

Many people who first encounter multidimensional beings are surprised when these beings or this intelligent awareness tells them that they are part of them, that they are one with them, and that their physical identity is only a small aspect of who they are.

These kinds of insights are very difficult to accept for a person who has spent their entire life believing that they are no more than a physical body. A direct encounter with your nonphysical friends can easily blow your mind. The fact that most people you know probably haven't had such an experience can make you fear you're going crazy.

It is easier to give up and accept that you are mentally ill than to question the most fundamental assumptions of our culture. But mental and social friction is only a transitional phase if one follows these encounters to their conclusion. Not many people dare to go all the way, but I believe that if you choose to stay true to your experiences they will lead you through your fears, through insanity, through the death of your false self and the rebirth of a new version of yourself. The process of re-identifying with your essence can be scary, but it is infinitely rewarding.

Within our collective psyche, there is a field of fear that prevents us from claiming our eternal nature. When we have experiences with this field, it can often feel like you are alone against a massive cultural pattern set in motion thousands of years ago.

In the past, thousands of "witches" were hanged, burned, or tortured because they bypassed the authority of the priests and communicated directly with the "spirit world." The fear of being prosecuted for following the guidance of one's own mystical experiences was a central element of the power structure of Western culture. The Church didn't want a decentralized system to emerge, which is why it was vital that people believed that one could only communicate with the spirit world under terms defined by the Church. By prosecuting witches, the Church made sure that everyone was afraid of the direct experience of the spirit world.

If there really is such a thing as the collective unconscious, it would make sense that these fears still affect us today. When a person has "supernatural" experiences, they have to deal with how all of humanity has related to that kind of phenomena.

Fear is an exclusive consciousness modality. Fear-driven actions create separation and exclusion. When you are afraid of sudden changes in perception, you try to push your experience away and end up struggling against your own self-actualization. As Patrice Somé would say, you are preventing yourself from becoming aligned with the spirit that wants to merge with you. I interpret his terminology of "merging" and "aligning with powers from the other world" as a process in which different patterns of consciousness form harmonious relationships. Becoming aligned with a spirit is the same thing as integrating a new data set, or expanding your worldview. This expansion process is much harder if you are afraid, because fear tends to push one back to the known and away from the unknown.

There is an underlying process which expands our perception, and there is an interpretive model which accompanies this process. The interpretive model is like an interface which allows us to navigate the transition. One person might see a mountain spirit, while someone else might see aliens or ancestral spirits. Others might see only colors, yet others feel a presence. The form through which one connects to a transcendent intelligence varies, but if we are patient with our experiences, they transmit meaningful emotions and realizations, which are the essence of a process that transcends the appearance of form and the metaphorical models we create. Our awareness perceives data streams and creates models, but if we are engaged in a process that changes the very essence of our awareness, the models we create and perceive will never be able to adequately represent the process.

During my meditations I have interacted with nameless and formless nonphysical intelligences that rearranged the energy flow in my brain (or, at least, that was how I perceived the process). Names are artifacts of human language, so it makes sense that nonphysical intelligences do not have names comprehensible by humans. If you meet a nameless being, perhaps it evolved in a rule set that did not have human language or even language based on sound. Their reality is different than ours like the reality of a fish is different from the reality of a bird. But if we practice our ability to tune into the essence of our awareness, we can experience a connection with other forms of intelligence.

The essence of our consciousness is much more than the part we manifest as the human personality. The part of us that has a human name and a human story is driven by an awareness that doesn't have a name. The connection to other intelligences happens on a level of our awareness that's not limited to the human identity. This level cannot always be perceived when our attention is focused on the subject-object relationship of time and space-based perception, yet it is always there, and we can work to develop our awareness of this deeper part of our selves.

When we are looking at something, or when we are recognizing something, there is a subject and there is an object; there is us and the thing we are perceiving. To go beyond that mode of perception, we need to become aware of awareness itself. If we practice this through meditation, we find that there is an incredibly vast "place" within us that is completely beyond anything we can believe or think. When you try to approach the core of your being, you'll find that logical thinking and language are a result of that which we are trying to explore. Your awareness is part of a reality that goes beyond the construct of identity and thought.

Our connection to other realities happens in this "place." It is communication on a non-rational level, which fills us with insights which then have to be translated into human thoughts. When you have an experience like that, it doesn't provide you with scientific proof, but it does provide you with the conviction that you are much more than you thought you were. You will have the experiential knowledge of a multidimensional reality.

CHAPTER 36:

Ego and Soul

Finding balance between different aspects of our self

The challenge of integration is the challenge of translation. The personality we built in three dimensions isn't capable of understanding the reality of the soul, but there are methods and concepts that can build bridges and allow the ego to become an integrated expression of the soul.

Ego, *personality self*, and *small self* are all terms that refer to the part of us that navigates our experience of separate identity. The small self cannot function in a rule set in which there are no beginnings and no ends. However, during a rapid awakening, we can experience a dimension of reality in which everything is connected and nothing truly starts or stops. This doesn't make any sense to the small self, which is why it is difficult to bring back the gifts of transpersonal experiences. It is like trying to merge two huge data sets compiled by completely different programming languages.

Finding a common denominator is difficult because the small self has a hard time dealing with the unfamiliar. But the unfamiliar, or intellectually unknowable, seems to be a central element of transpersonal experiences. This is why the integration process can feel like a tug-of-war. One aspect of yourself is trying to find rational explanations, while another aspect is trying to surrender to the experience of all knowledge within no knowledge, to the experience of complete unity with the absence of rational approximations.

I have met many people who find a balance within the dichotomy of ego and soul by thinking of the ego as a servant. The ego wants to be the master, it wants to take the credit for the soul's creative energy, but it cannot understand the way the soul navigates a life. There is no way for me to know this with certainty, but it seems likely to me that the network of our souls is also responsible for the strange coincidences that often appear to be guiding our paths. The ego doesn't understand why we suddenly met a random person in the subway who ended up changing our life. The ego doesn't understand why we suddenly reconnected with an old friend who we haven't talked to in years.

So much about our lives seems to be random, but this randomness might be an expression of a larger intelligence which is utterly incomprehensible to the small self. We can align our ego with this larger pattern if we believe it to be an expression of superior intelligence. This is difficult to do, though, because from the standpoint of the ego it isn't justifiable: *Why should I surrender to something that seems completely random and isn't always aligned with my definition of good and bad?*

Though there might not be an intellectual answer to this question, transpersonal experiences can provide you with trust and belief in a benevolent intelligence. If these experiences are properly integrated, this trust can transform the ego's need to control the unfolding of the future, making the ego the servant of a transcendental will. With this attitude, things fall in place naturally, even if they aren't necessarily explainable.

CHAPTER 37:

Words From the Network

Encounters with another form of intelligence

Paul Selig used to be a professor at NYU and a massage therapist. As he worked with his clients, he began to hear words in his head, sometimes names, sometimes little phrases. After a while, he began to ask his clients about these names, and was surprised how heavily these words affected their emotional states. It turned out that the names that were coming to him were the names of people who triggered of some kind of emotional conflict within his clients. Gradually, Selig began to trust the voices he heard, and the communication increased until he was hearing enough to fill entire books.

Over the years, Selig has written several books by repeating the words he heard in his head. His books never require any editing, and he doesn't take any credit for them. He says that the guides have written them.

I don't know why some people hear the voices of their guides very clearly while others do not. It might be dependent on a soul's specific relationship to its psyche. Perhaps some people have guides that are more integrated in the form of their superegos, while others perceive that part of themselves as another entity. Or maybe it depends on what lessons a soul is trying to learn through its physical life or what it wants to accomplish here on earth.

I believe that the small self, the personality of a single life, is like the tip of an iceberg. It is a representative of a network that transcends time and space. Often people who have out-of-body experiences say that they are becoming one with several other beings, and often people who talk to their guides get the message that their guides are another aspect of themselves, which many refer to as their higher self, superego, or superconscious self.

It seems that all of us have a part within us that is wiser than other parts. Not everyone perceives that part to be talking to them directly, but I assume that all of us sense at one point or another that there's a discrepancy between the part of us that is wise and altruistic and the small self that is tightly focused on our simplest desires and fears.

Sigmund Freud conceptualized this with the relationship of ego and superego. He didn't believe, though, that the superego is a part of us that transcends time and space. He thought the superego was merely a psychological construct which internalized the expectations of our parents and the rest of our social environment.

This idea was later rejected by psychologists such as Abraham Maslow, who pointed out that the superego is not always aligned with the expectations of parents or social environments. For example, there are many people born into abusive and criminal families who go on to pursue a benevolent path. Although they were raised in a negative environment, they still heard a voice that guided them toward a life they had never experienced, but knew was possible.

I believe that our superego is our soul's interest to actualize itself, and I believe that interest is shared among many beings that do not have a body, or a time and space personality. I believe that the actualization of our soul is a collective project of a free will network that is moving towards higher coherence, or a complex expression of love consciousness.

I have had conversations with what I perceived to be my non-corporeal guides during which they told me that I did not exist and that the experience of individual identity is an illusion. Hearing this from them temporarily unraveled the construct of individual identity. I experienced sadness and pain in coming to the realization that I am a collective project which incorrectly believes himself to be an individual. It felt like my own death.

Trying to wrap our minds around the idea of collective identity can be very confusing, as it doesn't make much sense while we are in bodies that seem clearly separate from their environments. But if you think about yourself, you will find more and more components, different drives, fears, desires, ideas, goals, labels, beliefs, identifications and so on. The only thing about you that isn't dividable is your awareness, which is the part of you that is aware of whatever you are focusing on. This aspect of yourself is your essence and it doesn't need an identity to function. In fact, you can be more focused if you aren't distracted by identity.

On that level of pure awareness, we are no longer form-based creatures, and our reality is vastly different than what the causal logic of our human brain can conceptualize.

Neuroscientist and consciousness researcher John Lilly has researched consciousness with sensory deprivation chambers. Several decades of self-experimentation have led him to experience states of consciousness that are usually only talked about by mystics. According to Lilly, during one of his sensory-deprivation experiences he perceived himself as connected to many other beings that were all part of an eternal network.

From Lilly's book *Simulations of God*:

> *When one tunes into high energy communications networks in special states of consciousness, he is reassured by the then-existing fact that he is a node in such a network; that there is constant information being fed into him, being computed below his levels of awareness and transmitted to others. This is all done with extremely high energy, far above what one usually experiences while in ordinary states of consciousness in the body. One experiences streaming of energy from unknown sources; streaming of energy going towards unknown sinks. A few of the nearby "other nodes in the network" may be visible. In such states there is nobody, there is only pure streaming energy, carrying information. In such a state one suddenly realizes that he is far more than he assumed he was when in the body, and yet he is also far less in terms of ego. In this state, he is a cosmic computer - small size - connected into the rest of the cosmic computers and into a huge universal computer. During such experiences, one feels the connections between all these computers as love, respect, awe, reverence, curiosity, and interest. And yet there is a high degree of efficiency with which the traffic is handled in these information channels.*

The first time I came across the term "eternal network" was during my brief relationship with Julia. By this point, I was already used to the fact that she occasionally perceived herself to be speaking to nonphysical beings. She saw them as having eight heads that were fused in a circle. She said their heads looked very alien, a little bit like insects or turtles, but they weren't frightening because she had the sense that she was one of them or one with them. They always came to her when she needed help and transmitted a powerful feeling of peace and love.

One night she was lying next to me and said that she heard them talking to her. She repeated the words back to me and I asked her if she would mind if I recorded what she was saying.

"Let me ask," she said, then paused. "Ok, yes, you can record it."

I took out my phone and started recording.

She had her eyes closed. The street lights were shining through the curtains and her curly hair covered her purple pillow. She whispered, with a lot of pauses between the words.

"When the human brain is relaxed, more information can be digested and received. That is why it is important to focus and relax the thoughts. Stress inhibits neural pathways, blocking receptors and inhibiting function. Stress manifests in many forms. When the body becomes stressed, it cannot respond to external stimuli in an appropriate way. It is too preoccupied fighting an internal battle and therefore becomes much more susceptible to pathogens and negative energy. The body becomes like an empty house that can be inhabited by whoever."

She paused for a moment and continued, eyes still closed.

"In order to maintain a healthy system, one must forgo the societal norms. Walking down a path of fulfilling our altruistic tendencies. Only when you look into the world of synchronized patterns can you connect the sequence of latent truths. You can always see the patterns manifest in many ways, internal and external. Observe both, they are the same. You don't have to look far, everything is always right in front of you. You cannot find what you are looking for by thinking about it. That is why it is tricky."

I felt shivers through my entire body. I was hesitant to interrupt her, but I also wanted to understand what she was saying. "What do you mean by 'the sequence of latent truths'?" I asked.

"The sequence of latent truths. That means that there is a particular order in which things happen and that they are happening intentionally for a reason and that they are right in front of your eyes and that it is pure, which means it's true, you don't have to think about it too hard, it's like a plant. It's true and it reacts completely honestly to its environment. If the plant does not have sun it does not grow, if it does not have water it does not grow. It needs space and it needs the appropriate amount of elemental nutrition. Just like anything else, just like anything else, it is simple."

There was a long pause. Julia still had her eyes closed and was breathing slowly.

"What about a situation that makes you grow and stresses you at the same time?" I asked. "Or a situation in which you need to grow because it stresses you, when do you know when to pull away?"

"I cannot help you with that," came the response. "That is your own process. When the vine of the plant runs into the wall it takes a different path, it just knows. It's a feeling, you can't think about it. You will know when the times are right, it's a feeling and you have helpers, you'll know. Stress can be good, existing can be stressful, there is much stimuli, and that's why it's important to stay centered, be the still point, just be the still point, you can watch it, the world, just watch it."

"May I ask who I am talking to?" I queried.

"Why must you know?"

"I am a curious being."

"Very curious indeed. You are speaking to a being that you don't have to know about. There does not have to be a name. Why does there have to be a name?"

"What about all the theories in my book that are so rational, are they helpful?"

"Rational theories are very helpful and appropriate, you live in very rational times."

I asked a few personal questions following that exchange, but instead of providing answers, the voice came back with: "This is your process, I cannot help you with that. I was hoping that you would ask a question about something less related to yourself. You already know the answers relating to your personal matters."

"I am sure I'll have many questions tomorrow," I said. "I just can't think of any right now. Maybe there is a thing you want to tell me?"

"This is actually not just about you. Maybe it would be beneficial for you to go to a place in yourself where you are of service to the people who influence you the most. Maybe it would be helpful for you to focus on gratitude and serving. Serving is very important at this time, spread that. Serve each other, be part of the complex web, be a participant in the growth. Just how a mushroom creates a doorway for information to pass through, so must you serve the network. Be a clear conscious pattern. The clearer your conscious pattern is the more easily information can travel through your consciousness. The point is that this information is transmitting to those around you and serves the network. It is not about your personal issues, those are transitory, this is an eternal web, make sense of it, be part of it, be part of the sensible pattern, harmonize, less about personal issues, those are transitory, don't let those consume you, that is your biggest problem at this time, you think about that much too much, it is a waste of energy. Energy is a gift, you are only given a certain amount of energy, use it. Be virtuous, not in a heroic way, plants are virtuous. They naturally embody purity. They follow the cosmic laws, they don't need to be told what to do, they know, the patterns are already there, you just have to get out of the way, it can happen on its own. Be the still point. Does that help you?"

"Very much," I said. "Very kind of you."

"It is just the way."

Then the voice addressed my ex-girlfriend. "You also have some information to receive from me. You need to stop being so rigid. Your patterning is such that you cannot be so engulfed in material outcomes. That is distracting you. You don't need such a result all the time. You are doing good work already, why do you need an outcome, what is that? Why? It is already perfect. Just allow the cosmic patterns to become fully expressed within your consciousness. Allow it to be fully present. You can never go too far from where you have already been. Your path will present much resistance in the outdated modes of perception. You must know that this resistance is only temporary. You can always enter a state of supreme ecstasy if you just calm the environment in which you inhabit. Become a source of calm for the people in your life. You don't have to go far to get to where you have been. You can always come back to where you are from. But your work here is important, good job..."

Worried that I wouldn't be able to get this chance again, I jumped in. "I would like to talk to you more on a regular basis and write down your advice," I said.

"How often do you mean?"

"Maybe every two weeks."

"Oh... Time... Of course!"

This was the end of this experience. Julia was quiet for a while and I had goose bumps on my whole body.

The next time I spoke with these beings I asked them how many of them there were and if they had bodies somewhere. They said that they are eight beings and that they did not use bodies anymore, because they existed as frequencies. I asked them how eight of them could speak through the vocal cords of one human. They said that they were able to do so through unity vibration. When I asked what unity vibration was, they told me unity vibration extends through all that is. Then they told me to stop asking them about their reality, saying that I was like a fish trying to understand the life of a bird. They told me to be more focused on my immediate reality.

Then I thought about my other ex-girlfriend, who I was still heart-broken over. Julia suddenly opened her eyes and said: "They told me that you wanted to ask about your ex-girlfriend, but you didn't because you did not want to make me jealous."

This was exactly the thought process I had, and I felt a little freaked out.

My relationship with Julia made me think that there could be many more people who perceive themselves to communicating with nonphysical beings. For Julia, it was a very private thing that she didn't even share with her family. She felt no need to scientifically explore the phenomena or convince people who don't believe in mystical phenomena. Her spiritual growth and exploration was something she did only for herself.

The following summer, my relationship with Julia transitioned into a friendship, and during a surfing trip in Portugal I met Elena, who is now my wife. I made a point not to talk to her about the subjects of nonphysical beings because I wanted to get to know her better before revealing that part of myself, but I had the feeling that she was also familiar with a deeper reality.

The first night we spent together, I was lying in a hammock and asked her if she could tell me a story. Given that I hadn't yet brought up my interest in the topic, I was quite surprised at how Elena started her story.

"When I was 12 years old," she said, "my mom brought me to a ceremony with a shaman."

She continued. During the ceremony, Elena told me, the shaman asked her to get up and walk towards the woods. The shaman began banging his drum. As she walked silently into the dark, two huge beings floated towards her. By her description to me, they were about 9 feet tall and had the outline of human figures, but their physical features were obscured by the bright light that emanated from them. She said they looked as if they were covered by a glowing white blanket.

Elena had never seen anything like this and was shocked and scared. She ran back to the shaman and told him what had happened. He told her that she had entered another state of consciousness in which she was able to perceive nonphysical energies that most people cannot see. At 12 years old, though, such an explanation didn't mean much to her.

The following week she encountered the beings again. This time it was in the city and she yelled at them, telling them to leave her alone, as she didn't feel ready to encounter them. They immediately disappeared and she never saw them again.

That wasn't her only encounter with the rationally inexplicable, though. Since Elena had been 10 or 11 years old, she had been drawing figures with elongated skulls. It was years later that she discovered that these beings had been depicted throughout history by many cultures, including the ancient Egyptians and the Paracas, who lived about 2,500 years ago in what is now Peru. Archeologists excavating ruins in these and other areas have discovered elongated human skulls that are thousands of years old.

A year after we met, Elena told me she'd encountered the elongated skull beings in nonphysical form. Like me, she's a musician, and she had been working on a solo album. Feeling stuck in her creative process, she asked for help during one of her meditation sessions. Suddenly she saw two beings with elongated skulls descending from her room's ceiling. One of them handed her a glowing ball, which she perceived as a representation of "source energy," or the essence of life. Immediately afterward, she got up and wrote and recorded a song. Over the following months, she went on tour with her solo album and released the song that she wrote that night, "Landscapes."

When I asked her if she thinks that the completion of her solo album had anything to do with that experience, she said, "I don't put these kind of experiences into a rational relationship with what I do."

Neither Elena nor Julia think that it is necessary to define such experiences. For them, the value is in the experience itself and they don't care whether the experiences are explainable. Personally, I don't see it quite the same way. I believe that if our awareness really is part of a multidimensional network of nonphysical beings, then science should explore that. It would change the way we see the universe and change the way we look for answers and guidance.

CHAPTER 38:

Rational Models

The need for rational explanations and their limitations

When I first met the beings who told me that my entire world and personality wasn't real, that I was just processing a bunch of information for the purpose of collective evolution, I found myself disappointed and upset that no one had told me sooner.

At first, it was also a huge blow to my ego and sense of identity. The ego wants to be the ruler in the kingdom, and when he is told that he is just a tool to navigate space and time, it makes him feel very small and insignificant. Eventually, though, I realized I didn't have to be upset by the concept. My small self or ego has found peace with the idea that I am a servant of an eternal network. The ego or small self actually gets very unhappy when it is cut off from its purpose. If we are not in service of love, we become selfish, greedy, and frustrated beings who hate themselves but are too stubborn to admit it.

It seems as if our souls have a built-in mechanism that only produces complete fulfillment when we are in service of something larger than ourselves. If we have spent our lives only focused on our small selves and our biological needs, it can be a painful shock when we suddenly recognize a larger reality. But this ego death is part of an essential process, and many spiritual traditions emphasize this idea.

In esoteric Christianity, it is sometimes referred to as the "crucifixion of the small self." In shamanism it is also known as death and rebirth. The shaman's old personality dies and he is reconstructed in a new way. The Qigong teacher Robert Peng also talks about this process. Peng was instructed by his master in China to meditate for a month in a dark cave without any food. Robert Peng now refers to the transition he went through as "small death, great life."

During these types of ego death events, the person experiencing it believes himself to be dying. After the acceptance of death, there is suddenly a shift in consciousness, and the recognition of eternal life fills the body with new energy. This can actually feel like an inner explosion of extreme magnitude. If this new energy is properly integrated, everything goes back to normal, except now the small self feels that it is a servant of something much greater which it loves and is loved by.

The kind of fulfillment one experiences after giving oneself to the divine is far greater than anything that could have been created of one's own volition; suddenly you feel the currents of love that are flowing through everything. You feel aligned. You feel taken care of. You feel at peace.

I don't know if it is necessary to believe in eternal existence to become self-actualized. There are people who don't believe in anything, yet still carry the same feeling of love and peace. I don't think that there is one rule that applies to all.

I personally needed to conceptualize my experiences in order to integrate them. I needed an explanation for who or what nonphysical beings are and why I perceived them to be communicating with me. Before my experiences, I had no belief system that was capable of explaining the experience of nonphysical beings and I felt in desperate need of a theory that could explain this kind of experience. But perhaps I only needed a belief system because I prematurely entered a larger reality. There are experienced meditators who don't find it necessary to approximate metaphysical experiences with rational concepts. To them, all of it is a creation of the mind, and they aren't interested in dealing with that domain. Even the question of whether the mind exists beyond the physical body doesn't make a difference to them. The instruction of most Buddhist meditation techniques is to ignore all visions or extrasensory experiences. If you are good at meditating and your only interest is in reaching a place without thought, you don't need to have any conceptual framework or belief systems.

Belief systems are programs that are executed through thoughts. If you practice having no thoughts, you also practice having no beliefs. When you are completely without thought, you're not thinking "Ah, I am part of an eternal network of love, these are my guides." You might experience it directly, but it doesn't have any rational meaning to you, therefore you have no need to put it into these terms.

Rational frameworks, like the ones I have been exploring, are only useful for those people who still have the need to think. Others might find it completely pointless to create such complex theories to describe something that is so obvious to them. This is something Julia and Elena taught me. But many people still have the need for rational explanations since we live in a society that is based on the assumption that there is a material reality out there which can and should be understood with thoughts.

Thoughts and rational belief systems are powerful models that stir the direction of human progress, but it's important to remember that models are just tools for understanding reality, not reality itself. Models can help us make predictions and find relationships between events, but on the path of self-actualization, models gradually lose their importance. The direct experience becomes more enriching than the process of intellectualizing. It doesn't matter anymore if you are a Buddhist, a Christian, a Muslim, or if you are using a fringe theory based on the simulation hypothesis. Whatever framework you use to conceptualize nonphysical realities is ultimately just a framework. Eventually you will arrive at an experiential reality that you no longer need to explain to yourself. This state will be the death and rebirth of your small self.

The intellectual framework presented in this book isn't a static belief. It is a patchwork of hypotheses that attempt to give new meaning and new concepts to ancient truths and show their reemergence in the findings of today's science. The truths that have compelled the creation of all myths and religions will always try to be understood anew. As technology and science advance, so does our vocabulary, metaphorical understanding, and empirical evidence. There will always be a dynamic process that uncovers the same truth in different conceptual frameworks.

If we can remember that all intellectual ideas are approximations, we can avoid the fixation on names and metaphors that have created religious wars, oppressive dogmas, and obsessive beliefs. The nonphysical aspect of existence is necessarily beyond our human language, yet our awareness has emerged from it. At best, intellectual ideas can help us integrate personal experiences and scientific discoveries, but the truth always lies beyond the framework.